VALUATION, HEDGING AND SPECULATION IN COMPETITIVE ELECTRICITY MARKETS

A Fundamental Approach

THE KLUWER INTERNATIONAL SERIES IN ENGINEERING AND COMPUTER SCIENCE

Power Electronics and Power Systems

Series Editor
M. A. Pai

Other books in the series:

OPERATION OF RESTRUCTURED POWER SYSTEMS
Kankar Bhattacharya, Math H.J. Bollen and Jaap E. Daalder, ISBN 0-7923-7397-9
TRANSIENT STABILITY OF POWER SYSTEMS: *A Unified Approach to Assessment and Control*
Mania Pavella, Damien Ernst and Daniel Ruiz-Vega, ISBN 0-7923-7963-2
MAINTENANCE SCHEDULING IN RESTRUCTURED POWER SYSTEMS
M. Shahidehpour and M. Marwali, ISBN: 0-7923-7872-5
POWER SYSTEM OSCILLATIONS
Graham Rogers, ISBN: 0-7923-7712-5
STATE ESTIMATION IN ELECTRIC POWER SYSTEMS*: A Generalized Approach*
A. Monticelli, ISBN: 0-7923-8519-5
COMPUTATIONAL AUCTION MECHANISMS FOR RESTRUCTURED POWER INDUSTRY OPERATIONS
Gerald B. Sheblé, ISBN: 0-7923-8475-X
ANALYSIS OF SUBSYNCHRONOUS RESONANCE IN POWER SYSTEMS
K.R. Padiyar, ISBN: 0-7923-8319-2
POWER SYSTEMS RESTRUCTURING: *Engineering and Economics*
Marija Ilic, Francisco Galiana, and Lester Fink, ISBN: 0-7923-8163-7
CRYOGENIC OPERATION OF SILICON POWER DEVICES
Ranbir Singh and B. Jayant Baliga, ISBN: 0-7923-8157-2
VOLTAGE STABILITY OF ELECTRIC POWER SYSTEMS
Thierry Van Cutsem and Costas Vournas, ISBN: 0-7923-8139-4
AUTOMATIC LEARNING TECHNIQUES IN POWER SYSTEMS,
Louis A. Wehenkel, ISBN: 0-7923-8068-1
ENERGY FUNCTION ANALYSIS FOR POWER SYSTEM STABILITY
M. A. Pai, ISBN: 0-7923-9035-0
ELECTROMAGNETIC MODELLING OF POWER ELECTRONIC CONVERTERS
J. A. Ferreira, ISBN: 0-7923-9034-2
MODERN POWER SYSTEMS CONTROL AND OPERATION
A. S. Debs, ISBN: 0-89838-265-3
RELIABILITY ASSESSMENT OF LARGE ELECTRIC POWER SYSTEMS
R. Billington, R. N. Allan, ISBN: 0-89838-266-1
SPOT PRICING OF ELECTRICITY
F. C. Schweppe, M. C. Caramanis, R. D.Tabors, R. E. Bohn, ISBN: 0-89838-260-2
INDUSTRIAL ENERGY MANAGEMENT**:** *Principles and Applications*,
Giovanni Petrecca, ISBN: 0-7923-9305-8
THE FIELD ORIENTATION PRINCIPLE IN CONTROL OF INDUCTION MOTORS
Andrzej M. Trzynadlowski, ISBN: 0-7923-9420-8
FINITE ELEMENT ANALYSIS OF ELECTRICAL MACHINES
S. J. Salon, ISBN: 0-7923-9594-8

VALUATION, HEDGING AND SPECULATION IN COMPETITIVE ELECTRICITY MARKETS
A Fundamental Approach

by

Petter L. Skantze
Caminus Corporation

Marija D. Ilic
Massachusetts Institute of Technology

KLUWER ACADEMIC PUBLISHERS
Boston / Dordrecht / London

Distributors for North, Central and South America:
Kluwer Academic Publishers
101 Philip Drive
Assinippi Park
Norwell, Massachusetts 02061 USA
Telephone (781) 871-6600
Fax (781) 681-9045
E-Mail <kluwer@wkap.com>

Distributors for all other countries:
Kluwer Academic Publishers Group
Distribution Centre
Post Office Box 322
3300 AH Dordrecht, THE NETHERLANDS
Telephone 31 78 6392 392
Fax 31 78 6546 474
E-Mail <services@wkap.nl>

Electronic Services <http://www.wkap.nl>

Library of Congress Cataloging-in-Publication Data

A C.I.P. Catalogue record for this book is available
from the Library of Congress.

Printed on acid-free paper.

Printed in the United States of America

To my parents, Lisbeth and Bengt,
for their love and support.

P.L.S.

To my husband Jeff,
who continues to believe in what I do.

M.D.I.

Table of Contents

PREFACE .... **xii**

1 INTRODUCTION **1**

2 OVERVIEW OF VALUATION AND HEDGING THEORY **7**

2.1 VALUING A COMMITMENT OPTION 7
2.2 MARKET BASED VALUATION 12
2.2.1 Spot Forward Relationships in a Risk Neutral World *14*
2.2.2 Information Content in Forward Markets *15*
2.2.3 Spot Forward Relationship when Firms are Risk Averse *16*
2.2.4 Arbitrage Pricing Theory *17*
2.2.5 Application of Arbitrage Pricing Theory in Valuing Forward Contracts *19*
2.2.6 Model Based Pricing and Dynamic Replication *21*
2.2.7 Limits to Arbitrage Pricing Arguments *23*
2.2.8 Spot Forward Dynamics for Storable Commodities *24*
2.2.9 What is Term Structure Modeling? *26*
2.2.10 Physical Interpretation of Term Structure *27*

3 OVERVIEW OF THE COMPETITIVE ELECTRICITY INDUSTRY **31**

3.1 DESCRIPTION OF MARKET PARTICIPANTS 31
3.1.1 Generation Companies *31*
3.1.2 Load Serving Entities *31*
3.1.3 Power Marketers *32*
3.1.4 Exchanges and Market Makers *32*
3.1.5 Independent System Operators *33*
3.2 ELECTRICITY MARKETS 34
3.2.1 The Spot Market *34*
3.2.2 The Physical Forward Market *35*
3.2.3 The Financial Futures Market *36*
3.2.4 The Derivatives Markets *39*

3.2.5 Generation Assets and Non-Standard Contracts..............39

4 ARBITRAGE PRICING AND THE TEMPORAL RELATIONSHIP OF ELECTRICITY PRICES..................................41

4.1 IS ELECTRICITY REALLY NON-STORABLE?................................43

4.1.1 Storage Strategies in Oil and Gas Plants..........................43

4.1.2 Storage Strategies in Hydroelectric Dams........................46

4.2 ARBITRAGE AND THE RELATIONSHIP BETWEEN PHYSICAL AND FINANCIAL CONTRACTS FOR ELECTRICITY..49

4.2.1 Application of Arbitrage Pricing Theory in the Relative Pricing of Physical Forward and Financial Futures Contracts on Electricity..49

5 BUILDING A PRICE MODEL FOR ELECTRICITY MARKETS..53

5.1 STRUCTURE OF MODEL..54

5.2 MODELING APPROACHES..56

5.2.1 Quantitative Modeling of Electricity Prices 56

5.2.2 Production Based Modeling of Electricity Prices............57

5.2.3 Economic Equilibrium Models of Electricity Prices........58

5.2.4 Agent-based Modeling of Electricity Prices 58

5.2.5 Experimental Modeling of Electricity Prices 59

5.2.6 Fundamental Modeling of Electricity Prices....................59

6 A BID-BASED STOCHASTIC MODEL FOR ELECTRICITY PRICES 61

6.1 LOAD CHARACTERISTICS..61

6.2 SUPPLY CHARACTERISTICS..62

6.3 PRICE AS A FUNCTION OF LOAD AND SUPPLY..........................63

6.4 STOCHASTIC LOAD MODEL..66

6.4.1 Modeling Demand Seasonality...66

6.4.2 Modeling Load Uncertainty 68

6.4.3 Mean Reversion..70

6.4.4 Stochastic Growth 70

6.5 STOCHASTIC SUPPLY PROCESS 72

6.5.1 Seasonality of Supply *73*

6.5.2 Modeling Supply Uncertainty *74*

6.5.3 Modeling Unit Outages *75*

6.5.4 Modeling Scheduled Maintenance *78*

6.6 SUMMARY OF THE BID-BASED STOCHASTIC PRICE MODEL 79

6.7 CALIBRATION OF THE BID-BASED STOCHASTIC MODEL 79

6.7.1 Generating a Time History of Supply States *81*

6.7.2 Estimating Deterministic Seasonal Load and Supply Shapes *82*

6.7.3 Calibration of Principal Component Vectors *83*

6.7.4 Estimation of the Volatility, Drift and Mean Reversion Rate *87*

6.8 THE TIME-SCALE SEPARATED BID-BASED STOCHASTIC MODEL 92

6.9 SIMULATIONS 98

6.9.1 Monthly Parameters *100*

6.9.2 Properties of the Daily Weight Process *102*

6.9.3 Daily Price Simulations *105*

6.9.4 Daily Price Simulations with Time Scale Separation *106*

6.10 CONCLUDING REMARKS 110

7 OPTIMAL FUTURES MARKET STRATEGIES FOR ENERGY SERVICE PROVIDERS 113

7.1 HEDGING RISK FOR ENERGY SERVICE PROVIDERS 113

7.2 THE PHYSICAL AND ECONOMIC INTERACTION OF ENERGY SERVICE PROVIDERS AND THEIR CUSTOMERS 114

7.3 PROBLEM FORMULATION 116

7.4 MODELING 117

7.4.1 Cash Flow Model *118*

7.4.2 Price and Demand Models *120*

7.4.3 Modeling the Firm's Risk Preference *124*

7.4.4 Summary of the Hedging Problem *124*

7.5 EFFICIENT REFORMULATION OF COST FUNCTION 125

7.6 SOLUTION APPROACHES 126
7.7 THE END STATE PROBLEM 127
7.7.1 Static Optimization Over Multiple Delivery Periods 129
7.8 THOUGHTS ON THE COMPLEXITY OF THE ESP HEDGING PROBLEM 131

CHAPTER 8 VALUING GENERATION ASSETS 135

8.1 INTRODUCTION 135
8.2 A PRINCIPAL COMPONENT BASED PRICE MODEL FOR ELECTRICITY SPOT MARKETS 136
8.2.1 Formulating the Unit Commitment Decision 137
8.2.2 Price Model Used in the Unit Commitment Problem..... 139
8.3 CREATING A LOOKUP TABLE OF CASH FLOWS 140
8.3.1 Incorporating Stochastic Fuel Prices............................. 141
8.4 LINKING SIMULATED PRICES TO THE LOOKUP TABLE TO GENERATE SIMULATED CASH FLOWS 143
8.4.1 Generation Asset Valuation.. 143
8.5 CONCLUDING REMARKS 145
8.6 FIGURES 146

9 MODELING LOCATIONAL PRICE DIFFERENCES 149

9.1 INTRODUCTION 149
9.2 LOCATIONAL PRICING AND MARKETS FOR TRANSMISSION ... 149
9.2.1 Pricing Flows in Electric Power Networks 149
9.2.2 Contracts for Transmission 150
9.2.3 Valuing Transmission Rights.. 151
9.3 MODELING TRANSMISSION RIGHTS AS A DERIVATIVE ON SPOT PRICES 154
9.4 OVERVIEW OF EXISTING PRICE MODELS 157
9.5 INTERACTIONS BETWEEN NEIGHBORING MARKETS 158
9.6 VALUING A TRANSMISSION RIGHT 161
9.7 SIMULATION BASED VALUATION 161
9.8 DYNAMIC HEDGING 165
9.8.1 Dynamic Hedging and the Bid-based Model.................. 166

9.8.2 Implications on the Valuation of a Spread Option *167*
9.8.3 Replicating a Flexible Transmission Right under the Bid-based Price Model *169*
9.9 GENERALIZATION OF THE MODEL TO A 3 NODE EXAMPLE 171

10 INVESTMENT DYNAMICS AND LONG TERM PRICE TRENDS IN COMPETITIVE ELECTRICITY MARKETS 177

10.1 INTRODUCTION 177
10.2 A LONG TERM MODEL FOR ELECTRICITY PRICES 178
10.2.1 Stochastic Demand Process *178*
10.3 MODELING INVESTMENT DYNAMICS 179
10.3.1 Backward Looking Investment *179*
10.3.2 Forward Looking Investment *182*
10.4 A DYNAMIC NOTION OF RELIABILITY 185
10.5 EFFECTS OF GOVERNMENT POLICY 187
10.5.1 Comments on Simulation Results *189*
10.6 CONCLUDING REMARKS 190

11 CONCLUSION 193

APPENDIX A 197

APPENDIX B 201

REFERENCES 207

INDEX 213

Preface

The challenges currently facing participants in competitive electricity markets are unique and staggering: unprecedented price volatility, a crippling lack of historical market data on which to test new modeling approaches, and a continuously changing regulatory structure. Meeting these challenges will require the knowledge and experience of both the engineering and finance communities. Yet the two communities continue to largely ignore each other. The finance community believes that engineering models are too detailed and complex to be practically applicable in the fast changing market environment. Engineers counter that the finance models are merely statistical regressions, lacking the necessary structure to capture the true dynamic properties of complex power systems. While both views have merit, neither group has by themselves been able to produce effective tools for meeting industry challenges.

The goal of this book is to convey the fundamental differences between electricity and other traded commodities, and the impact these differences have on valuation, hedging and operational decisions made by market participants. The optimization problems associated with these decisions are formulated in the context of the market realities of today's power industry, including a lack of liquidity on forward and options markets, limited availability of historical data, and constantly changing regulatory structures. To address these challenges we reevaluated key premises underlying modern finance theory, including the role of arbitrage pricing theory (APT) in markets for non-storable commodities. Without the ability to store an underlying commodity, traders cannot create risk free arbitrage portfolios to take advantage of spot and forward price differentials. As a result, one of the basic framework used in finance theory to model the joint dynamics of spot and forward prices does not work for electricity markets. However, rather than reject the approach taken in traditional market models, we sought to expand existing models to incorporate underlying physical and economic processes that drive the dynamics of supply and demand in the marketplace. In doing so we gained access to new data from which the distribution of future price paths could be implied. While it is intuitive that the dynamic properties of physical events such as temperature changes and forced generation outages directly effect the term structure of electricity prices, the challenge is to create a model that is simple enough from a calibration and optimization standpoint, without loosing the unique properties which each of these fundamental processes contribute to the market. From this modeling paradigm emerged the Bid-Based Electricity Price Model, introduced in Chapter 6, the core of this book.

To members of the energy industry reading this book, it should be mentioned that while the model is presented with algorithms for calibration to market data and applications in hedging and valuation problems, it is not meant as a ready-made tool for solving company specific trading and risk management problems. Rather it is an attempt to explore a richer set of modeling and optimization tools to address the unique challenges of competitive electricity markets. We hope that it will inspire you to do the same within your organization.

Acknowledgments

This book is written by two engineers whose interest in the dynamics of large scale power systems led them to dig deep into the fundamental assumptions of finance theory, and its application to the power industry. A critical portion of this journey was two summers spent by the first author working in the industry, the first summer with a developer of software tools for the energy industry, the second on the trading floor for a leading power marketer learning the details of the trading and risk management operation. These experiences provided a wealth of open questions, which were brought back to the university within the context of MIT's Energy Laboratory Consortium on competitive power systems. The first author pursued his doctoral degree at MIT under the supervision of the second author, and this book is an outgrowth of their joint efforts. Both authors greatly benefited from a continuous interaction with the members of the consortium, consisting of industry participants ranging from trading organizations to software vendors to regulatory agencies. Two representatives of this group who stand out due to their relentless input of new questions and ideas are Ralph Masiello of Caminus Corporation, and Scott Weinstein of Constellation Power Source. Equally important was the input from the vibrant academic community at MIT. Through their critiques and insights, professors John Tsitsiklis and Richard Larson were instrumental in linking the research to related problems in the areas of optimization and stochastic modeling. Finally, the students in the competitive power systems group, especially Andrej Gubina and Michael Wagner who worked closely with the authors on parts of this research, helped provide a unique environment for learning and exploration. The authors would like to thank them, and all who have directly or indirectly contributed to this work.

Chapter 1

Introduction

The purpose of this book is to build a framework to solve physical and financial commitment decisions contingent on electricity, in a deregulated market environment. The set of problems include valuing investment opportunities in physical assets, valuing and hedging obligations to serve customers, and the pricing of electricity dependent derivative contracts. Electricity markets suffer from a severe case of over-dimensionality. Due to the lack of economic storage of the commodity, each time interval of delivery can be considered a separate product. Furthermore, the scarcity and complexity of the transmission system leads to significant locational variations in price. The combination of the temporal and spatial properties of electricity poses significant barriers to market liquidity, and makes it exceedingly hard for market participants to solve valuation and risk management related optimization problems.

A key step in overcoming the curse of this dimensionality problem is the development of effective dynamic models of the uncertainty in the underlying asset price. Model based valuation and hedging formulation take advantage of the distinct relationships in the market in order to reduce the complexity of the space of possible future price combinations. This approach allows the user to form dynamic replicating portfolios, and leads to computationally efficient implementations of various optimization problems. The tradeoff is that if the relationships specified in the model do not hold, then the user may drastically underestimate his risk exposure. This additional component, known as model risk, is important to keep in mind when reading this book.

There are many approaches to building financial models. Specifically we distinguish between statistical models, based on regressions or best fits between different data sets, and fundamental models, consisting of postulates and data, together with a means of drawing dynamical inferences from them [1]. Choosing an approach is contingent on the nature of the traded asset, as well as on the availability of market data. Statistical models generally require a rich history of market prices for calibration and testing purposes. Fundamental modeling works best when there are strong physical or economic relationships embedded in the market, which can be translated into model constraints.

Electricity is a natural candidate for the fundamental modeling approach. Markets are still emerging, so there is little price history available. Furthermore, constantly changing regulatory structures invalidate much of the existing historical data. On the other hand, the demand and supply of electricity are governed by a set of exogenous drivers, including weather patterns, economic growth, and fuel cost. In addition, the transmission of electricity is governed by Kirchhoff's current and voltage laws, describing the relationship between power injections at various nodes, and the flow on the transmission lines. One of the main contributions of this book is the construction of a stochastic model for electricity price, founded on the fundamental drivers and constraints outlined above, and applying of this model to temporal and locational optimization problems under uncertainties.

Inter-temporal uncertainty in storable commodity markets is a well-researched area in finance. Of specific interest is the correlation between the spot price and the price of forward contracts with varying maturity dates, also known as the term structure of the market. The ability to store a commodity provides a natural link between the spot and forward price. If the forward price increases, a producer may sell forward contracts, remove his product from the spot market, and put his production output in storage to deliver against the forward contracts. This creates a supply deficit on the spot market, driving prices up until they are back in equilibrium with the forward market. Similarly, if forward prices drop, market participants will empty their storage facilities onto the spot market, purchase forward contracts, and refill the storage facilities at a later date by taking delivery on the forward contracts. This creates a downward pressure on the spot price, again pushing it back into balance with the forward price. The most important lesson from the simple qualitative examples described above is that markets for storable commodities have strong inter-temporal links, governed by spot and forward price signals. Arbitrage arguments impose bounds on sustainable relative price levels in the market. The lower the cost of storage, the tighter the bounds, and the stronger the inter-temporal price correlation. In Chapter 2, we introduce examples of stochastic models for storable commodities as well as stocks (which have zero storage cost), which are currently used in the marketplace. It is interesting to note that these models all use spot price as a state variable. This makes sense, since inter-temporal production and storage decisions are directed by price signals. Additional states can include the convenience yield, describing a market preference at the time of consumption of the commodity.

It is tempting simply to adapt the existing models for commodity prices to electricity markets, with minor cosmetic changes. Indeed many power marketers still use versions of oil and gas models for their electricity operations. The problem is that the underlying assumptions behind these models, based on storage arguments, do not hold for electricity. Electric power cannot be simply purchased, stored and resold at a later date in response to price signals from the spot and forward markets. It has been argued that the ability to store the fuel used to produce electricity (oil, gas, coal and water) is equivalent to the ability to store the commodity itself. This approach is incomplete, since it does not take into account constraints on the rate at which the fuel can be converted into electricity. In fact observations of actual intra-day price behavior indicate that market participants are to a large extent unable to execute temporal arbitrage to a significant extent, even on short time scales. As a result the inter-temporal link between spot and forward markets, or spot prices in different time periods, is no longer governed by a set of price signals as was the case for storable commodities. This does not mean that electricity markets do not have a well defined term structure. It does indicate however that the key to understanding the term structure is not to be found in price centered arguments.

One of the contributions of this book is to construct a dynamic model for the uncertainty in electricity prices, by relating the price changes to a set of underlying physical and financial processes. The underlying states relate directly to the demand and supply processes in the market, and spot and forward prices are modeled as outputs of these processes. While seemingly a modest transformation, the change in the set of state variables has tremendous implications, such as:

1. We are better able to explain the complex term structure of electricity prices. The model illustrates how each fundamental input contributes to the inter-temporal correlation of electricity prices. Furthermore we are able to capture the complex, multi-timescale seasonality prevalent in electricity markets.
2. New sources of information become available in the calibration of the price process. One can explicitly use the regional demand history to calibrate the model, providing a longer and more stable source of data for the emerging industry.
3. The model links the price of electricity to other traded commodities and contracts such as fuel (oil, gas, and coal), and weather derivatives (rainfall and temperature). This provides new opportunities for hedging electricity risk, or creating cross commodity arbitrage portfolios.
4. There is potential to use the model in combination with economic game theory to price derivative contracts in emerging markets with no price history.
5. The models naturally couple price and quantity risk, providing an efficient method for hedging fixed priced commitments to serve customers (standard offer contracts).

The second aspect of electricity which makes it a unique commodity is its complex transportation structure. Electric power is transmitted over a meshed network of transmission lines, and cannot be controlled to follow point to point paths. The non-linear relationship between power injections at the nodes, and the flows on the lines, are governed by Kirchhoff's current and voltage laws. As a result of this physical phenomenon, the locational pricing of electricity is more involved than that of most other commodities. The spot prices at neighboring markets can diverge drastically if the connecting line is congested. This is of particular concern to generators and loads who have signed commitments to deliver power across frequently congested interfaces. This has led to the emergence of a number of physical and financial transmission contracts, to be used by market participants to hedge locational risk. There still, however, does not exist a consistent method for pricing these contracts. Specifically, there seems to be no clear relationship between the price of the inter-nodal transmission contract, and the forward and options contract prices at the respective nodes. One approach is to use traditional spread option models. These models generally assume that the two underlying assets each evolve as random walk processes, and that any interaction can be captured by estimating the correlation between the Wiener processes. In the book we question the applicability of traditional models in describing the stochastic properties of price spreads between electricity spot prices at different network locations.

Using a model based on supply and demand states, one can explicitly incorporate the network flow constraints. The resulting probability distribution of the price spread is in stark contrast to what is implied by traditional approaches. This has tremendous implications for the pricing of transmission rights and locational price derivatives. The model is also extended to allow for market based analysis of hypothetical expansions to the transmission grid, allowing transmission companies to estimate the value of new investments.

The final section of the book covers the dynamics of new investment in generation capacity. It is shown how the time delay in information from the spot market to the investment decision, and further in the installation process of new power plants, can lead to periods of over and under capacity. We further analyze the relationship between price trends and the physical reliability of the market. Specifically we address the dangers related to excessive government intervention in order to curb price spikes.

Chapter 2

Overview of Valuation and Hedging Theory

This book will address a number of decision problems faced by the participants in competitive power markets. The set of problem formulations can best be summed up as commitment problems, since they involve some form of physical or financial commitment from the firm in question. They include investment in physical assets, agreements to deliver electricity to customers over a specified period, bilateral financial agreements, and numerous exchange traded derivatives based on electricity prices.

In this chapter we show how such decision problems can be treated as instances of a general optimization formulation. We begin by examining the existing literature in finance covering investment and valuation procedures, starting with the process by which an individual firm values a given commitment option. Next we address how multiple firms with different objectives interact through centralized markets, leading to the concept of a market valuation. Finally we see how, under rigorous market assumptions, the two forms of valuation can converge. The theoretical basis of the convergence is dictated by arbitrage pricing theory (APT). We analyze APT in a static and dynamic framework, illustrating its applications in equity, interest rates and commodity markets.

2.1 VALUING A COMMITMENT OPTION

The traditional approach to valuing a commitment option is the net present value (NPV) rule. To calculate the net present value of a commitment (C), one first characterizes the distribution of the payoff ψ_t for every time t over the horizon of the commitment $[t_o, T]$. The payoff incorporates any cost and revenue associated with the commitment, and can be positive or negative at any point in time. The net present value is calculated by integrating the payoff, adjusted by a discount rate δ, over the horizon of the commitment.

$$NPV_{t_0}(C) = \int_{t_0}^{T} e^{-\delta(t-t_0)} E_{t_0}\{\psi_t\} dt$$

where E{.} is the expected value operator, which is necessary since future payoffs are generally uncertain.

The NPV rule states that if the net present value is positive, then the firm should enter into the commitment, assuming there are no other commitment options available. If there are multiple options, the firm should choose the one with the highest NPV. To use the NPV criteria, one must overcome two challenges. The first is estimating the expected value of future payoffs. The second challenge lies in determining the appropriate discount rate δ. The discount rate contains information about the time value of money, or the opportunity cost of any capital which is tied up in the commitment and cannot be spent elsewhere. This cost would generally be set equal to the firms cost of raising new capital, for example by issuing bonds. For the remainder of the book, unless stated otherwise, we are going to assume that this component of the discount rate is equal to zero. As long as the cost of capital is deterministic, this assumption can be made without loss of generality.

The discount rate must also reflect the level of risk in the investment. Assuming the firm is risk averse, the discount rate will increase with the level of uncertainty associated with future payoffs. A commitment option with lower expected cash flows, may be preferable to an option with higher expected cash flows if the risk level is significantly lower. Intuitively this argument is easy to understand. Quantifying a firm's risk preference is more difficult. One possibility is to equate the level of risk with the variance of the cash flow. This puts a greater requirement on the modeling of future cash flows, since the firm needs to estimate variances in addition to expected values. Furthermore, a firm needs to consider the risk associated with the cumulative cash flow, not just the instantaneous variance at each point in time. To calculate the variance of the sum of the cash flows, the firm needs to estimate the covariance matrix of all the future cash flows.

$$\psi^{tot} = \sum_{t=t_o}^{T} \psi_t$$

$$\mathrm{var}\left(\psi^{tot}\right) = \sum_{i=t_o}^{T} \sum_{j=t_o}^{T} \mathrm{cov}(\psi_i \psi_j)$$

The firm can then define its risk preference by stating its utility (U) in terms of the tradeoff between the expectation and variance of the return on a commitment. An example of this is the mean variance utility function

$$U_t\left(\psi^{tot}\right) = E_t\{\psi^{tot}\} - r * \mathrm{var}_t\{\psi^{tot}\}.$$

For a given commitment option, the r parameter of the mean variance formulation can be mapped uniquely to the discount rate δ in the NPV formulation. However, two projects with the same variance on the total payoff may end up with different discount rates depending on the timing of the cash flows.

A second formulation which is popular among decision makers is the value at risk (VAR) criterion. VAR estimates the amount of the firm's capital which is at risk of being lost during a given time interval. Capital is defined to be 'at risk' if the probability of a loss is greater than a threshold set by management. For a threshold probability X, we can define the value at risk as

$$\mathrm{Prob}\left(\psi^{tot} \leq -\mathrm{VAR}\right) = X$$

The value at risk formulation generally provides a hard constraint to optimization problems. For instance, a decision maker may wish to optimize the expected payoff from a set of possible commitment opportunities with associated payoffs ψ^i, given that it cannot exceed a maximum VAR.

$$\begin{aligned} &\mathrm{Max}_i\left(E\{\psi^i\}\right) \\ \text{s.t.}\quad &\mathrm{Prob}\left(\psi^{tot} \leq -\mathrm{VAR}\right) \leq X \end{aligned}$$

This technique provides a probability estimate of the worst case scenario, which makes it intuitive for managers to use. However, it is computationally harder to apply since we need higher order information about the joint probability distribution of the payoff. Specifically the value at risk is very sensitive to high impact low probability events, which create 'fat tails' in the distribution of payoffs.

The above discussion indicates that the analysis leading up to a decision rule for the firm can be broken into two parts. First, the firm must estimate the stochastic properties of future payoffs associated with the commitment. Second, the firm must characterize its own risk preference in relation to the proposed investment. Once these have been established, the firm can attempt to solve the optimization problem and arrive at a decision rule. This process is outlined in Figure 2-1.

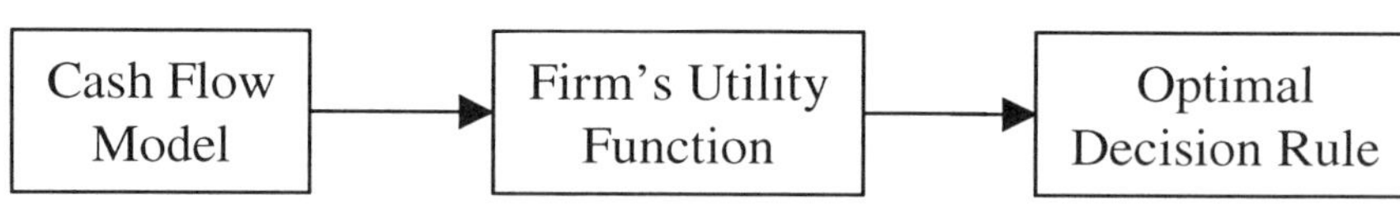

Figure 2-1: The Firm's Decision Process

The process described in Figure 2-1 covers a wide range of applications. We will now focus on a subset of these where the commitment is based on an asset traded in an organized market place. The market functions so that for each instance in time it sets a price for the traded asset. It is further assumed that the uncertainty in cash flow from the commitment is only dependent on the uncertainty of future price levels in the market. The decision process can then be amended as shown in Figure 2-2.

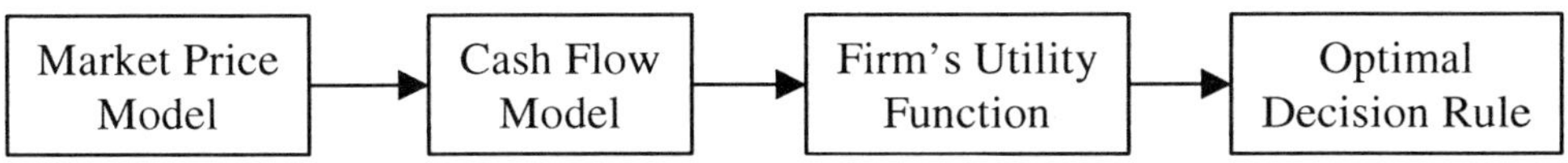

Figure 2-2: Amended Decision Process

As shown, Figure 2-2 makes an implicit assumption about the relationship between the market price and the firm's decision rule. The price model is taken as an exogenous input to the decision rule, implying that the firm's actions will have no effect on the market. The market, however, is simply a collection of firms, each with its own decision rule. The ability of a firm to influence the market price is often referred to as the firm's market power. Depending on the number and size of participants, the decoupling of the market price from the individual decision process may or may not be prudent. We will discuss this assumption in detail in the chapter on market modeling approaches. For now we recognize that a more general formulation of the decision problem would include a feedback loop from the decision rule to the price model as shown in Figure 2-3.

The existence of a market, and a market price, has a fundamental impact on the way we think about the commitment problem. It combines the preferences of all participating firms, and arrives at a single number for the value of an asset. The value assigned to an asset by the market may be substantially different from that seen by the individual firm. In fact it is the discrepancy between the way in which different firms value assets which allow markets to exist, and add value, in the first place. In the following section we examine the information contained in market prices, and how they can be used by individual decision makers.

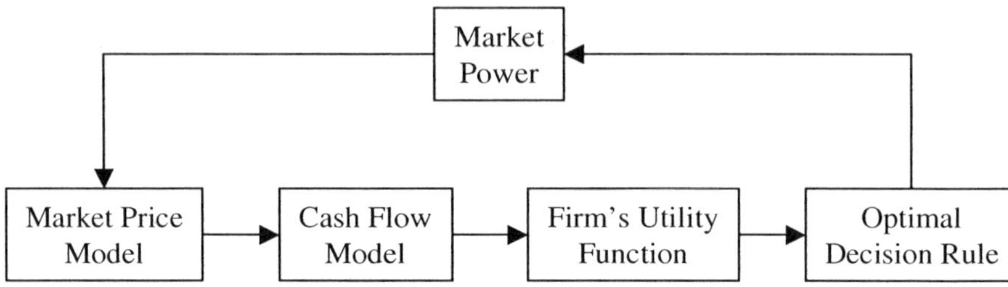

Figure 2-3: Incorporating Market Power into the Decision Process

2.2 MARKET BASED VALUATION

So far we have assumed that the market assigns a value to the traded asset at any given point in time. This value, known as the spot price, is the cost of purchasing the asset for immediate delivery. The spot market, however, is only one of several venues available for a firm to trade in a given asset. Mature markets generally trade forward contracts as well as numerous options contracts in addition to the spot. Forward contracts specify the delivery of a given quantity (q) of the asset for a fixed price (F) at a specific future time (T), known as the maturity of the contract. The seller of the contract generally has the option of paying the prevailing spot price at maturity rather than delivering the actual asset. The payoff for the buyer, or long position, of the forward contract at maturity can therefore be written as

$$Payoff\,(forward, long) = q\left(S_T - F\right).$$

The cost of purchasing a forward contract at any point in time t prior to maturity is denoted by F(t,T). The asset whose spot price determines the payoff of the contract is known as the underlying asset.

There are numerous variations of options contracts traded on equity and commodity markets. One the most common is the European call option contract, which gives the buyer the option to purchase the underlying asset at a fixed strike price (X) at the maturity of the contract. The payoff for the long position of a European call option is

$$\text{Payoff(call, long)} = \max\{0, (q(S_T - X))\}.$$

The price of purchasing the contract at a time t prior to maturity is denoted by $c^E(t,T)$.

The payoff, and therefore the value, of the forward and call option contracts are functions of the spot price of the underlying commodity. We refer to the general category of traded contracts with this property as derivatives of the underlying asset. It is important to note that the commitment for the individual firm, as defined in the previous section, is effectively a derivatives contract. The payoff from the commitment is a function of future levels of the spot price. All commitment opportunities discussed in this book are effectively derivatives of the electricity market, or the price of affiliated commodities, some with multiple underlying assets. A key distinction is that some derivatives, such as forward contracts, are publicly traded, while others are available to only a limited set of investors, at a given point in time. In this section we address the relationship between traded and non traded contracts. Particular emphasis is put on the information value of price signals from traded contracts, and to which extent they can be used to imply the value of non-traded contracts. A distinction has to be made between the value of a contract to an individual firm, and the market valuation of the contract. The value of a contract to a firm is contingent on the firm's risk preference. The market value of a traded contract on the other hand is uniquely defined by its current price.

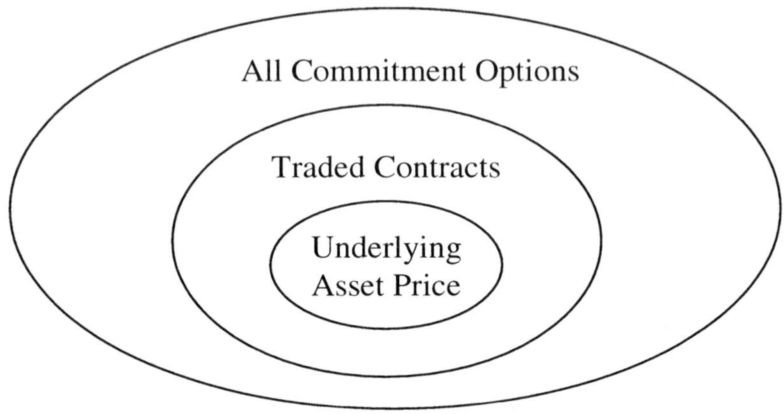

Figure 2-4: Space of Possible Commitment Options

First we examine the relationship between the various traded contracts, and the underlying asset's price. Consider first the relationship between the spot and forward prices.

2.2.1 Spot Forward Relationships in a Risk Neutral World

By studying the historical behavior of the spot market, one might arrive at a reasonable model of the stochastic properties of future spot prices. Assume that, at time t, a firm uses all available historical information about the spot market to estimate the distribution of the spot price at some future time T. Next the firm observes a forward price F(t,T). We assume that participants in the forward market are rational. That is, they will purchase or sell a forward contract only if it increases their total utility. For the sake of discussion, we assume that all participants have risk preferences which can be characterized by the mean variance formulation, so that each individual firm j, has a utility U^j defined by the properties of the total cash flow ψ^j.

$$U_t^j(\psi_T) = E_t\{\psi_T^j\} - r^j * \mathrm{var}_t\{\psi_T^j\}$$

Furthermore, assume that all cash flows are the result of trades on the forward market, and all cash flows occur at time T. If the utility function of all participants were independent of risk, that is, the r parameter were zero for all firms, then the utility from owning a long position in the forward contract is given by

$$U_t^j\left(\psi_T^{long}\right) = E_t\left\{S_T - F(t,T)\right\} = E_t\left\{S_T\right\} - F(t,T).$$

Similarly, the utility of owning a short position in the forward contract is given by

$$U_t^j\left(\psi_T^{short}\right) = F(t,T) - E_t\left\{S_T\right\}.$$

Furthermore, since the expectation operator is linear, the marginal change in utility from purchasing or selling a forward contract is independent on the other contracts held by the firm, even if they are dependent on the spot price S_T. Assuming a rational market in which market participants will not engage in trades with a negative utility, the only possible forward price at which trades could occur, is given by

$$F(t,T) = E_t\left\{S_T\right\}.$$

2.2.2 Information Content in Forward Markets

The above statement regarding the price of a forward contract in a risk neutral environment, makes a subtle but powerful assumption about the role of information in markets. The expression $E_t\{S_T\}$ states that we are taking the expectation of the spot price at time T, given a set of information available at time t. This assumes that the same set of information is available to all market participants, or equivalently, that all information in the market is public. We can relax this assumption by indexing the expectation operator to each firm j, $E_t^j\{S_T\}$, allowing for private information to be held by a subset of market participants. The question then arises to which extent the resulting forward price reflects the presence of private information. A firm which holds confidential information that the spot price will be higher than anticipated may trade in sufficient quantities to raise the forward price. But will the forward price reflect the entire price increase anticipated by the private information, or will it be a weighted average of the public and private valuation? These types of questions regarding the informational content of forward markets are still unresolved in the finance community. We bring them up here since the role of information flow will be crucial in developing and calibrating models for the spot and forward prices in later chapters.

2.2.3 Spot Forward Relationship when Firms are Risk Averse

We now reconsider the pricing of a forward contract assuming that firms are risk averse, so that the r parameter in the mean-variance formulation is greater than zero. This initially leads to a contradiction. Consider two firms with risk averse parameters r_1 and r_2 respectively. If both firms are rational, the following set of constraints must hold in order for the two firms to enter into a forward contract, with firm 1 taking the long position:

$$U^1 = E_t\{S_T\} - F(t,T) - r_1 \operatorname{var}\{S_T\} \geq 0$$
$$U^2 = F(t,T) - E_t\{S_T\} - r_1 \operatorname{var}\{S_T\} \geq 0$$
$$r_1, r_2 \geq 0 .$$

There is no set of risk premium r^i which will satisfy this criteria. Does this mean that forward markets cannot exist if all market participants are risk averse? What we have neglected is that when the variance of the return is incorporated into the utility measure, the incremental change in the utility is no longer independent of the firm's other trades. Specifically, if a firm has an existing obligation with payoffs which are negatively correlated to the spot price at time T, then entering into the long position in a forward contract will tend to reduce the overall variance of the firm's returns. The forward contract then takes on the function of a hedge, offsetting the risk from an underlying position. In commodity markets, a major function of forward markets is to hedge the positions of producers and consumers of the commodity. Producers have physical assets which make them naturally long in the commodity (with payoffs positively correlated to the spot price), while consumers are naturally short (with payoffs negatively correlated to the spot price). When a producer sells a forward contract to a consumer, both parties decrease the overall variance of their future cash flows, and it is therefore possible to find a set of forward prices for which both sides increase their expected utility.

The above example illustrates an important aspect of derivative pricing. When risk premiums are nonzero, the firm must consider its entire portfolio of existing obligations when valuing a new contract. Furthermore, it should consider any combination of available derivative contracts. This not only makes it challenging for the firm to model its decision problems, but it also makes it extremely difficult to speculate on the relative prices of spot and forward markets. To determine the price of forward contracts from a bottom up perspective, one would need knowledge of each firm's risk preference, as well as all other contracts held by the firm at any given time. Modeling derivative price from this perspective is clearly a gigantic task, even if the necessary information were to be available. There are other approaches to characterizing the relationship between various traded contracts. One approach centers on building portfolios out of available contracts so that the uncertainties cancel each other out, leaving the holder with a guaranteed return. The existence of such portfolios limits the combination of prices which rational firms would tolerate in the market place. In the next section we review the theory behind this approach, which is known as arbitrage pricing theory.

2.2.4 Arbitrage Pricing Theory

Arbitrage pricing theory (APT) [2] is based on the assumption that a market cannot sustain price levels which allow market participants to construct portfolios which the possibility of a positive return with no risk of loss and no initial investment. Such trades are known as pure arbitrage opportunities. This assumption imposes constraints on the manner in which prices evolve in the market. We adopt the following definition of arbitrage. Consider a market with n tradable assets, each with price $x_i(t)$. A portfolio Π is built by purchasing and selling these contracts. The value of the portfolio is given by

$$\Pi(t) = \sum_{i=1}^{n} w_i(t) x_i(t)$$

where w_i represents the quantity of asset i in the portfolio. $w_i(t)$'s can be negative if the market allows short-selling. Since future asset values are uncertain, the value of the portfolio at $t>t_0$ is a random variable.

We define an arbitrage opportunity as follows. Arbitrage exists if at time t_0 we can construct a portfolio Π with the following properties:

$$\Pi(t_0) = 0$$
$$\text{and for some } t > t_0$$
$$\text{Prob}\big(\Pi(t) < 0\big) = 0$$
$$\text{Prob}\big(\Pi(t) > 0\big) > 0$$

This means that we can construct a portfolio with zero cost, which has zero probability of decreasing in value and a strictly positive probability of increasing in value. Since the portfolio has zero initial cost, any market participant can purchase an unlimited amount of the portfolio, and enjoy a risk free guaranteed profit. The theory is that as arbiters start to take advantage of this opportunity, they will create an upward price pressure on assets with positive weights in the arbitrage portfolio, and downward price pressure on assets with negative weights. Prices will then reach a new equilibrium where the arbitrage opportunity no longer exists.

2.2.5 Application of Arbitrage Pricing Theory in Valuing Forward Contracts

The creation of arbitrage portfolios is inherently linked to the temporal properties of the underlying asset. Of particular importance is the ability of the investor to store the asset.

We apply arbitrage pricing theory to three types of assets -- equity, storable commodities and electricity -- to illustrate how the characteristics of the underlying asset changes the constraints imposed on the relative prices of spot and forward contracts.

2.2.5.1.1 The price of a forward contract on a stock

Assume the current price of the stock, which pays no dividends, is S_t and the risk free interest rate is r, continuously compounded. The price of a forward contract on the stock (F(t,T)) with delivery date T must then be $e^{r(T-t)}S_t$. To see why this is true consider the following cases:

1. If $F(t,T) > e^{r(T-t)}S_t$, the investor should sell one forward contract, borrow S_t dollars at the risk free rate (assuming this is possible), and buy one unit of stock. The net cash flow at time t is zero. At time T, the investor delivers the stock against the forward contract, receives F(t,T) dollars as payment for the forward, and $e^{r(T-t)}S_t$ dollars to pay off his debt. The net cash flow at time T is $F(t,T)-e^{r(T-t)}S_t>0$. This is a pure arbitrage opportunity, which cannot be sustained in an efficient market, and therefore sets the upper limit for the forward price.
2. If $F(t,T) < e^{r(T-t)}S_t$, the investor should buy a forward contract, short-sell one stock, and lend S_t at the risk free rate. The net cash flow at time t is zero. At time T, the investor pays F(t,T) and receives delivery of the stock from the forward contract. He uses this stock to repay his short-selling obligation. He also recovers $e^{r(T-t)}S_t$ from the money loan. The net cash flow is $e^{r(T-t)}S_t-F(t,T)>0$. This is again a pure arbitrage opportunity, setting the lower limit for the forward price.

In this case the upper and lower limits for the forward price are identical and, therefore, in an efficient market where participants can borrow and lend at the risk free rate, the forward price must be given by: $F(t,T)=e^{r(T-t)}S_t$. This illustrates two important points. First, under no-arbitrage conditions, the forward price of a stock is a deterministic function of the spot price and the time to maturity (T-t). Second, there is a smooth convergence of the spot and forward prices at maturity.

2.2.5.2 The price of a forward contract on a storable commodity

Assume the current unit price of the commodity is S_t, the present value of the total cost of storage incurred during the length of the futures contract is U, and the risk free interest rate is r. The lower bound on the futures price for delivery at time T is $F(t,T)>(S_t+U)e^{r(T-t)}$. If this does not hold, an investor can receive a risk-free profit by borrowing S_t+U at the risk free rate, purchase the commodity and pay off the storage cost, and short a forward contract in the commodity. The cash-flow at time t is zero, and the cash-flow at time T is $F(t,T)-(S_t+U)e^{r(T-t)}>0$. This is known as cash and carry arbitrage.
Payoff at each time step from cash and carry arbitrage:

	t	T
Buy commodity to be delivered against forward contract.	$-S_t$	0
Sell forward contract	0	F(t,T)
Pay storage cost	-U	0
Borrow now, repay at maturity	S_t+U	$-(S_t+U)e^{r(T-t)}$
Total Cash Flow	**0**	$\mathbf{F(t,T) -(S_t+U)e^{r(T-t)}>0}$

Table 2-1: Cash and Carry Arbitrage

Cash and carry arbitrage establishes an upper bound for the forward price of the commodity. The bound converges to the spot price as we reach maturity (T=t), and hence if the forward price is lower than the spot price then the two prices must converge at maturity.

2.2.6 Model Based Pricing and Dynamic Replication

Much of the work in finance in the last thirty years is a direct outgrowth of a set of seminal papers published by Fisher Black, Myron Scholes and Robert Merton in the early seventies (see [2],[4]). In their work, Black and Scholes proposed that stock prices (S) could be modeled as a simple stochastic process known as geometric Brownian motion:

$$dS = \mu S dt + \sigma S dz$$

where dz is a non-differentiable, continuous time Wiener process. Next they considered a derivative f, contingent on S. By applying Ito's lemma, it can be shown that the price of the derivative contract must follow the following stochastic process:

$$df = \left(\frac{\partial f}{\partial S} \mu S + \frac{\partial f}{\partial t} + \frac{1}{2} \frac{\partial^2 f}{\partial S^2} \sigma^2 S^2 \right) dt + \frac{\partial f}{\partial S} \sigma S dz .$$

Next they recognized that since the same random input (dz) was driving both stochastic processes, an investor could hold a combination of the stock and derivative so that the random components cancel each other out perfectly. Specifically, the risk neutral portfolio is formed by purchasing one derivative contract, and short selling $\partial f/\partial S$ contracts of the stock. The process describing the change in value (or return) of the portfolio (Π) is then given by

$$d\Pi = -f + \frac{\partial f}{\partial S} S = \left(-\frac{\partial f}{\partial S} - \frac{1}{2} \frac{\partial^2 f}{\partial S^2} \sigma^2 S^2 \right) dt .$$

This equation does not contain any random terms. If the portfolio return is deterministic, Black and Scholes argued, the return must be equal to that of a risk free loan, i.e. the risk free interest rate. Equating the return of the portfolio to that from lending money at the risk free rate,

$$d\Pi = r\Pi dt$$

led to the Black-Scholes-Merton differential equation

$$\frac{\partial f}{\partial t} + rS\frac{\partial f}{\partial S} + \frac{1}{2}\sigma^2 S^2 \frac{\partial^2 f}{\partial S^2} dt = rf \ .$$

The price of the derivative can be found by solving this differential equation for a set of given boundary conditions, specific to that derivative. A European call option with strike price X and maturity date T, for example, will have the boundary condition given by

$$f_T = \max(S_T - X) \, .$$

Another use of the above relationships is to find a portfolio of stocks and bonds (assumed to be a risk free loan) which perfectly replicates the payoff of the derivative. This is known as a replicating portfolio. The replicating portfolio for a derivative f is given by holding ∂f/∂S stock contracts, and f-(S*∂f/∂S) dollars worth of risk free bonds. The ratio of the stocks to derivatives contracts is known as the delta of the derivative:

$$\Delta_f = \frac{\partial f}{\partial S} \, .$$

Another interpretation of the same line of reasoning is that if an investor can always eliminate the financial risk associated with holding a risky derivative by creating a replicating portfolio, then speculators will be unable to charge a premium for taking on this risk. As a result the market will be risk neutral. That is, the value of each derivative will be equal to the expected value of the discounted cash flow. This leads to a complementary approach to valuing derivatives known as risk neutral pricing.

Risk neutral pricing can be interpreted as the price which would prevail if market participants were ambivalent towards risk, or if the risk were spread out over so many participants that its effect on each individual firm became negligible. This is a much stronger assumption than that of no arbitrage. Investors are generally risk averse, and there are only a limited number of market participants holding any given asset. There are, however, assumptions under which the two pricing schemes converge, as in the case presented above. In general, if a perfect replicating portfolio can be formed, then risk neutral and arbitrage pricing will produce the same result.

2.2.7 Limits to Arbitrage Pricing Arguments

While APT provides a convincing argument for why physical and financial forward prices must be equal at all times, actual observations in the market place show that the two markets can diverge at times. The reasons for this inconsistency can be found in the assumptions underlying the arbitrage argument. The following points illustrate how market realities deviate from the theory:

1. **Moving Equilibrium:** Arbitrage pricing theory is based on an equilibrium argument. It states that in a market with active arbitruers, a set of prices which allow for risk-free profit with zero investment is not sustainable. As traders execute the arbitrage, they gradually alter the relative prices until the system settles into an arbitrage free state. Markets in general, and electricity markets in particular, are continuously evolving dynamic systems. The effect is similar to that of a feedback control system driving the states of a system towards a continuously changing control input. Unless the input signal evolves at a significantly slower rate, the states will never settle to their equilibrium value.

In the case of electricity markets, the validity of the equilibrium argument will depend on two factors:

1. The rate at which new information about the future expected value of the spot price enters the market. Changes in the traders' perception of the

future is the driving input in the futures market. Information which would cause traders to change their perception would include updates on future weather/load conditions, or news of a generator or transmission line outage.

2. The volume and rate at which contracts are trading in the market. This represents the magnitude and speed of the feedback response, or the rate at which the market can react to the new information. This is also known as a market's liquidity.

We address this issue in more detail as we introduce our dynamic model for the evolution of the spot price.

2. **Uniqueness of Prices:** Unlike the spot market, the forward markets do not have a unique clearing price. The price quoted by the exchange is a weighted average of all trades in the last day. However there is no guarantee that the trader can find a counter-party willing to trade at exactly this average price at the time the arbitrage is executed. There generally is a gap between the highest price the market is willing to buy, and the lowest price the market is willing to sell at. This is known as the bid-ask spread. The magnitude of the bid-ask spread is dependant on the liquidity of the market.

3. **Transaction Costs:** Exchanges are generally for profit enterprises. They make a profit by charging a small fee for every contract which is executed on the exchange. The loss incurred by the trader due to such fees is known as a transaction cost. In electricity markets, exchanges generally charge a fixed fee per MWh of power covered in the contract. In order to execute an arbitrage, the guaranteed profit must be greater than the total transaction cost incurred. The magnitude of the transaction cost is relatively minor. Nordpool for example charges approximately one cent per MW traded in a futures contract.

2.2.8 Spot Forward Dynamics for Storable Commodities

The effects of cash and carry arbitrage can also be interpreted as a dynamic relationship between spot and forward prices. Assume that at time t we observe a forward price F(t,T), which violates the upper bound imposed by APT. We would expect the following behavior in the market.

1. In the spot market, demand will increase, as arbitreurs rush to buy the commodity in order to store it. This puts upward pressure on the spot market price.
2. On the forward market, the same arbitreurs sell forward contracts in order to execute the arbitrage, creating downward pressure on the forward price.

Now consider the reverse condition, when forward prices drop below spot market levels. In this case, no pure arbitrage strategy is present, since it may not be possible to short sell a physical commodity on the spot market. However, consider the position of a market participant who is currently holding an inventory of the commodity. For this person, the optimal strategy will be to sell the inventory today, and purchase cheap forward contracts which can be used to restore the inventory at a later date. If there is significant inventory in the market, this will put downward pressure on the spot price, and upward pressure on the forward price.

One can question whether the bounds set by APT are valid under realistic market conditions. This is especially true for commodities with thin forward markets and high transaction cost. However, whether or not the bounds are quantitatively accurate, the qualitative interaction between spot and forward prices can certainly be observed.

1. An increase/decrease in the forward price will put upward/downward pressure on the spot price.
2. A spike/drop in the spot price will put upward/downward pressure on the forward price.

Consider the following scenario. Tomorrow OPEC announces that it will reduce its annual production of oil by 50%. Based on this news, the forward price of oil increases sharply. Next, arbitreurs recognize the disparity between spot and forward prices, igniting a buying spree on the spot market. This causes an immediate spike in the spot price of oil. The above scenario illustrates an interesting characteristic of storable commodities markets. The relationship between the state of physical production and consumption on one hand, and the spot market price on the other, is non-causal. In other words, a future drop in production leads to a spike in today's spot price. Note, however, that the relationship between the spot price and the information flow is still causal. That is, the spot market will only react when the drop in future production becomes known to market participants.

The need to model the dynamic relationship between the spot and forward prices in storable commodities has led to the notion of convenience yield (y) (see [2], [5], and [6]), which is defined as:

$$F(t,T) = (S_t + U)e^{(r-y)(T-t)}$$

The convenience yield represents the premium the market is willing to pay in order to physically hold the commodity today, rather than a promise for delivery at time T. We can model y as a deterministic parameter, or a stochastic state of the system depending on the market.

2.2.9 What is Term Structure Modeling?

One way to think of term structure is the dynamics of changes of the forward curve over time. It is important to realize that while we can model the spot price as a one dimensional stochastic process, the forward curve encompasses a number (N) of values, representing the current forward prices at each delivery time. To model the dynamics of the forward curve, one must therefore describe the joint evolution of these N values over time. This problem inherently has a much higher dimensionality than spot modeling. To handle the dimensionality problem, one can turn to basis functions. An intelligent choice of basis functions can reduce the dimensionality of the problem with a minimal loss in information. It can also make the model more intuitive. One common approach is to define basis functions so that one has a strong emphasis on the front end of the forward curve (contracts with a short time to maturity) , and the other has a strong emphasis on the back end (long time to maturity). Each basis function is associated with a random input, with an associated volatility. When traders refer to long and short vol., they are talking about the volatility of the input associated with the long and short basis function respectively.

2.2.10 Physical Interpretation of Term Structure

Spot prices in most commodity market tend to exhibit mean reversion. This refers to the tendencies of prices to revert back to long term equilibrium levels after temporary spikes. This behavior is inherently linked to the dynamics of the forward curve in these markets. Consider for a moment a highly competitive market with no risk premium, so that forward prices represent the market expectation of future spot prices:

$$F(t,T) = E_t\{S_T\}\,.$$

Mean reversion dictates that if there is a spike in today's spot price, the effect of that spike on the expectation of future spot prices will decrease with time (T). Equivalently, the effect of a spike in the spot price on the forward price F(t,T) is a decreasing function of the time to maturity (T-t).

Term structure is easiest to understand in the context of a specific model. One of the simplest, and most popular, is the one-factor, lognormal, mean reverting model

$$\frac{dS}{S} = \alpha\left(\ln\left(\frac{\mu}{S}\right)\right)dt + \sigma dZ \ ,$$

where S is the spot price, and Z is a Wiener process.

Applying the risk neutral assumption we arrive at the following model for the dynamics of the forward price:

$$\frac{dF}{F} = \sigma e^{-\alpha(T-t)} dZ \ .$$

We characterized the dynamics of the entire forward curve (theoretically infinitely far into the future, although one factor models should not be used beyond a limited horizon).

Note that although we arrived at the result from a different path, it is essentially equivalent to assigning a single basis function to the N dimensional space of forward prices.

A reduced order term structure model is a powerful tool for traders. It allows them to trade off risk across contracts with different maturities. If one believes that forward prices behave according to (2), then one can use a single type of forward contract to hedge perfectly all price risk over any future time horizon. The problem of course is that the model in (1) and (2) is a gross oversimplification of reality. The most obvious flaw is that the expectation of a future spot price does not change until one comes close in time. Equivalently, the back end of the forward curve never moves! This is a direct result of the assumption that all price movements are temporary. In reality this is not true. Changes in the available technology on the demand or supply side, evolving regulatory structure, or changes in the overall economic outlook, can all have long term effects on the commodity price, and will therefore impact the back end of the forward curve. Furthermore such events may have a high impact on the expectation of future price levels but have little or no effect on today's spot price.

In terms of the model, long term price uncertainty can be incorporated by adding a second stochastic factor relating to uncertainty in the mean. Such a model may take on the form

$$\frac{dS}{S} = \alpha\left(\ln\left(\frac{\mu}{S}\right)\right)dt + \sigma^S dZ^S$$

$$\frac{d\mu}{\mu} = \kappa dt + \sigma^L dZ^L \ .$$

This leads to a term structure model with two basis functions:

$$\frac{dF(T,t)}{F(T,t)} = e^{-\alpha(T-t)}\sigma^S dZ^S + (1 - e^{-\alpha(T-t)})\sigma^L dZ^L .$$

Chapter 3

Overview of the Competitive Electricity Industry

3.1 DESCRIPTION OF MARKET PARTICIPANTS

After the deregulation of the utilities industry, a number of new market participants have appeared in place of the old vertically integrated utilities. In this section five categories of market participants are introduced, covering the most crucial functions related to wholesale electricity markets: generation companies, load serving entities, power marketers, exchanges and market makers, and independent system operators. These categories are not mutually exclusive. Large generation companies often serve as load serving entities, and may also fill power marketing functions. Independent system operators and power exchanges, on the other hand, tend not to have other interests or investments in the market, since this would lead to conflicts of interest.

3.1.1 Generation Companies (GenCos)

In the deregulated marketplace, generation company refers to any firm which owns physical generation assets. These firms vary from large deregulated subsidiaries of the old utilities to single plant independent power producers (IPPs). In this book we will treat generation companies, as well as load serving entities and power marketers, as for profit entities, whose objective is to maximize profit.

3.1.2 Load Serving Entities (LSEs)

As of today there are no functioning retail markets for electricity. End users are supplied either through bilateral agreements, or out of the wholesale market. Since buying electricity wholesale involves significant transaction costs, customers are generally served by intermediaries known as load serving entities, or energy service providers. LSEs serve as aggregators, taking on a large number of residential, commercial and/or industrial customers. The functions of the load serving entities include estimating the aggregate demand of their customers and entering bids to the wholesale market in order to secure delivery of the estimated load.

3.1.3 Power Marketers

With deregulation came a significant increase in the financial risk of both generation companies and load serving entities. Many firms were not equipped to handle the risk management aspect of their electricity supply or delivery operation. This led to the emergence of a new class of market participants known as power marketers. There are two fundamental components to this type of firm's operation, a marketing and a trading section. The marketing section will approach GenCos, LSEs, or even large end users directly, offering to take on part of their electricity risk exposure at a premium. If an agreement can be worked out, the marketers will pass on the acquired risk to the trading section, which is responsible for hedging the power marketer's risk exposure. The reason a power marketer can get away with charging a premium for this service is twofold. It generally has an efficient trading operation, allowing it to minimize transaction costs and, more importantly, it has a good understanding of the source and correlation of the risk in the various contracts it takes on, allowing it to hedge out effectively most of the uncertainty through futures and options trading.

3.1.4 Exchanges and Market Makers

In the regulated industry, utilities generally supplied electricity to a clearly defined geographic area, using a set of native generators to fulfill the demand requirement. As a result there was little incentive to enter into contracts with neighboring utilities unless there was a critical power shortage. Deregulation has dramatically increased the number of transactions between different market participants. Generation companies will supply their energy to the highest bidder, independent of the location, and load serving entities are willing to import power if it will reduce the cost of serving their customers. To accommodate this increase in market activity, a number of exchanges have emerged for the electricity industry. Exchanges match buyers and sellers of electricity, and charge a small transaction fee for their services. The exchanges differ in their time frame of delivery (short term spot markets vs. longer term forward markets), and in trading in physical or financial commitments. A description of the types of markets available for market participants is provided in the next section.

3.1.5 Independent System Operators

The market entities described so far are generally for profit companies, looking to take advantage of the opportunities provided by the competitive marketplace. Electricity markets however, are delicate physical networks, which can easily break down if pushed beyond their operating limits. To ensure the physical safety of the grid, regulators have encouraged the formation of independent system operators (ISOs). An ISO is a non profit entity, which acts as a supervisor of the physical transactions registered between power suppliers and customers. The two main functions of the system operator are to balance power, and manage congestion on the grid. The power balancing problem is an inherent result of the non-storability of electricity, forcing the system operator to maintain a constant stand-by reserve of spinning power capacity, so that a sudden loss of generation on the system does not lead to a drastic drop in frequency. The congestion management problem results from a nonlinear relationship between power injections and flows, forcing the system operator to implement complex pricing systems to prevent the overloading of transmission lines (see chapter on multi market modeling).

3.2 ELECTRICITY MARKETS

There are three fundamental markets available for trading electricity: the spot market (day ahead), the physical forward or bilateral market, and the financial futures market. In addition, there are a number of standard as well as over the counter options contracts traded, either through exchanges or on a purely bilateral basis. Before attempting to develop models that describe the pricing of these various contracts, we need to understand the manner in which electricity is traded.

3.2.1 The Spot Market

Spot power is traded under a number of different market structures in the United States, ranging from power pools to power exchanges to independent system operators. The common ground among these markets is that they all involve a centralized auction mechanism to allocate which generating units should be used to meet the demand. In this section we describe the rules governing a typical power exchange. While rules may vary somewhat based on geographical location, this should serve as a good example to understand the decision process facing producers in the deregulated marketplace.

A producer wishing to sell power submits a bid curve to the exchange. The bid curve describes the willingness of the producer to deliver power as a function of market price. For example a producer may be willing to supply a total of 50MW if the price is \$20/MW, and may offer to supply a total of 100MW if the price increases to \$30/MW. Bid curves are generally supplied on a day-ahead basis, and a different bid curve may be specified for each of the 24 operating hours.

The exchange gathers all the bids from power producers, and similar bids from consumers. The bids are used to compile aggregate supply and demand curves for each hour. The intersection of the demand and supply curves determines the market clearing price (MCP). All supply bids with a price less than the MCP are accepted, and are paid the clearing price. Similarly all demand bids with a price higher than the MCP are accepted, and are charged the clearing price. This ensures that demand and supply commitments match perfectly, and also that the exchange remains revenue neutral.

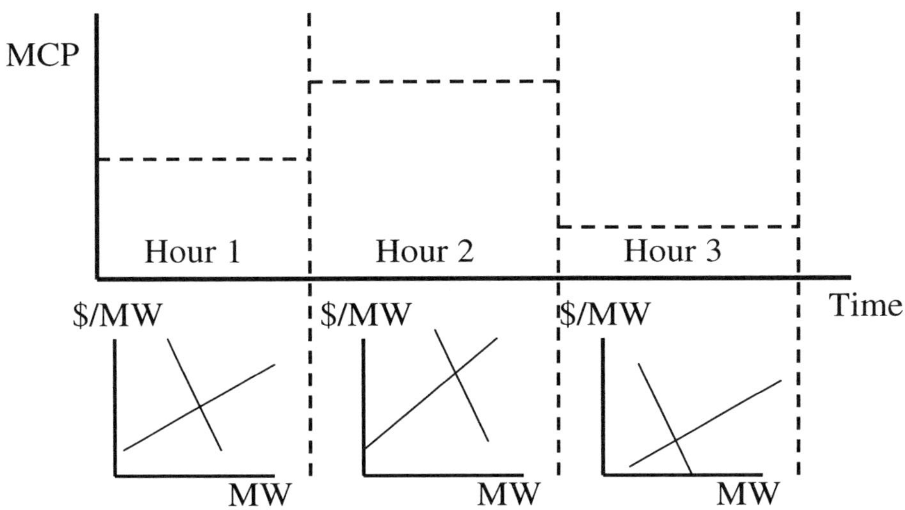

Figure 3-1: Spot Market Dynamics

3.2.2 The Physical Forward Market

Physical forwards can be traded on an exchange or in a bilateral manner through over the counter (OTC) transactions. Exchange traded forwards use standardized contracts, with power being traded in monthly on- and off-peak blocks. The contract specifies a single MW quantity (q) and a single $/MWh price (F). The short position (seller of the forward contract) is obligated to deliver power physically at a constant rate q to a location specified in the contract (the HUB). The contract does not specify the location at which the power is produced or consumed, but states that the short party is responsible for delivering the power from the generator location to the HUB, and that the long position is responsible for delivering the power from the HUB to the load location. For both sides this may involve purchasing additional transmission contracts, or purchasing/selling power through the spot market. Such provisions are not addressed in the contract, and the relative prices of the spot and transmission markets will not affect the price of the forward contract.

The price of exchange traded physical forwards is quoted daily by the exchange. The information provided includes the high and low prices as well as the volume traded and the volume of open interest. The exchange quotes prices for every delivery month up to 15 months into the future. This vector of prices G(t), which constantly evolves as new trades become public, constitutes the forward curve for electricity.

$$G(t) = \begin{bmatrix} g_{jan00}(t) \\ g_{feb00}(t) \\ \vdots \\ g_{mar01}(t) \end{bmatrix}$$

Physical forward contracts trade continuously while the exchange is open, until the fourth business day prior to the first delivery day of the contract. At this point trading terminates, and any party left with a short position is required to deliver power according to the provisions in the contract. A trader can avoid this by ‘booking out’ his position, purchasing a long position which exactly offsets his short position for the same delivery month.

3.2.3 The Financial Futures Market

Financial futures contracts for electricity are traded on exchanges such as the New York Mercantile Exchange (NYMEX) and the Chicago Board of Trade (CBOT). Financial contracts are similar to exchange traded physical contracts in structure. The main difference is that the parties entering into the contract have no intention of physically producing or consuming the power, but rather use it as a financial hedge against other positions in the market. The financial futures contracts are therefore settled by the exchange of cash rather than power. In general the payoff function for a party holding the long position in a forward contract is given by,

$$payoff(long) = S_T - F(t,T),$$

where S_T is the spot price at maturity T, and F(t,T) is the price of the futures contract at the time t it was entered into. The problem which occurs with electricity is that the delivery period for the futures contract is one month, while the underlying spot process is updated on a day-ahead basis. As a result, when the futures contract matures on the 4th business day prior to the 1st day of the delivery period, the spot prices for hours in the delivery month are not yet known. Hence the contract cannot be settled financially at this time. To circumvent this problem, exchanges have taken on two different approaches: ex-post settling and ex-ante settling.

Ex-post settling: In this approach, the futures contract is settled gradually during the delivery month. If two parties have entered into a futures contract for q MWs of on-peak power at a price F, then for every day for the duration of the delivery period, the following process determines the cash flow:

1. The on-peak price of power for the day is calculated by averaging the hourly price of the 16 on-peak hours from the day-ahead spot market. We denote this price as S^{peak}
2. The long position (buyer of the contract) will pay the difference between the S^{peak} and F times the quantity of the contract, times the number of on-peak hours (16). If this quantity is negative then the cash flow will be from the short position to the long position.

The total cash flow for the long position over the duration of the delivery month is given by

$$CF_{ex-post} = \sum_{i=1}^{n} 16q\left(S_i^{peak} - F\right)$$

where n is the total number of days in the delivery period.

Ex-ante settling: In this case, the futures contract is settled financially at its expiration date, i.e. on the 4th day prior to the beginning of the delivery period. Since the day-ahead spot price is not yet known for the delivery month, the price of a physical forward contract for the same delivery period and location is used in place of the day-ahead spot. This effectively is a change in the underlying commodity from which the futures contract's value is derived. The contract goes from being a derivative on the spot price to a derivative on the physical forward price. The payoff function for the long position at maturity T is given by

$$CF_{ex-ante} = \sum_{i=1}^{n} q\left(G(T,T) - F\right)$$

where q is the quantity of the contract in MWs, G(T,T) is the price of a physical forward on the last day of trading, and F is the price at which the futures contract was purchased.

Both the day-ahead spot and physical forward are based on the same commodity: electricity delivered at a specific grid location. However there is no simple mapping between the ex-post average spot price and the ex-ante physical forward price. This is a very crucial point to understand in electricity markets. While the settling procedure differs from market to market, the dominant trend seems to be in the direction of ex-post settling, as seen in California and Nordpool. Unless otherwise specified we will from now on assume that financial forwards settle ex-post.

3.2.4 The Derivatives Markets

A number of options and other derivative contracts are traded in the electricity marketplace. They can generally be grouped into three categories: temporal, locational, and inter-commodity derivatives. Temporal derivatives are the most common, and are used to hedge against future movements in the spot price of power at a given location. They include simple call and put options, as well as more sophisticated swing options and look-back (reference) options. Locational derivatives are generally used to hedge against the risk of volatile price spreads between the power production and delivery locations. These types of derivatives are closely linked to the trade of physical and financial transmission rights and will be addressed in detail in Chapter 9. Inter-commodity derivatives are contracts based on the price differential between electricity and another commodity. The spark spread for example is the spread between gas and electricity prices at a given conversion factor. The emergence of such contracts are a direct result of the significant dependence of electricity demand and supply on other commodities. The marginal cost of a gas fired generator, for example, is directly proportional to the gas price. Similarly, the strong correlation between electricity demand and temperature has resulted in a significant interest of power marketers in the emerging weather derivatives markets. We will return to the interaction between electricity and other commodity and derivatives markets in the modeling section of the paper.

3.2.5 Generation Assets and Non-Standard Contracts

In addition to options contracts, market participants need to be able to hedge returns from a number of assets and bilateral contracts. These problems can be considered as extensions of the derivative pricing problem. For example we can view a generator as a physical option to convert gas to electricity at a given efficiency rate. The financial equivalent of owning a generator would then be to hold a spread option between gas and electricity. In reality there are physical constraints on the generator which makes the payoff more complex, but the point is that we assign a value to a physical asset by modeling it as a single, or portfolio of, options contracts. We address this problem further in the section on dynamic replication.

Chapter 4

Arbitrage Pricing and the Temporal Relationship of Electricity Prices

The key issue in relating electricity spot and forward prices is storability. The lack of economic storage opportunities for electricity makes it impossible to form a cash and carry type arbitrage portfolio. As a result we cannot impose any arbitrage free bounds on the relative levels of spot and forward prices. It does not end there however. There is no limit on how far the prices of two forward contracts with different delivery months can diverge, since no arbitrage portfolio can be created to exploit the price differential. The story gets worse when we address the spot market. For a given year there are 8760 delivery hours, each with a unique price. There is no constraint on the relative price of spot power from one hour to the next. Relying purely on arbitrage theory, the number of random variables needed to define the spot and forward markets for one year would therefore be 8772. To use the forward market with a hedge we would need to estimate the entries into the 8772 by 8772 covariance matrix. Electricity markets suffer from a severe case of the curse of dimensionality. To circumvent this problem we must rely on effective modeling solutions. In the next chapter a detailed model for the spot market will be presented, describing the temporal correlation of prices as a function of fundamental drives. Before going through this process, however, we examine the relationship governing the spot and forward markets in the context of non-storability.

As stated earlier, temporal arbitrage is not possible in electricity markets due to the lack of storage. As a result, the dynamic relationship between the spot and forward prices described above does not hold for electricity. A good example is the case of scheduled unit outages. If it were announced today that a major nuclear plant in New England would be out of commission for the month of July, this would cause an immediate increase in today's price of a forward contract with delivery in July. However it would have no effect on the current spot price. We can therefore state that electricity spot prices are causal in the state of production and the consumption of electricity. This will have a tremendous impact on how we model electricity spot and forward prices.

Without the ability to execute an arbitrage between the spot and forward markets, APT is useless in predicting the relationship between the two markets. Instead we have to address the forces underlying the supply and demand in forward markets. One approach is to assume that the market as a whole is liquid enough that every participant holds a small fraction of the total risk. As a result the market effectively behaves in a risk neutral manner, even if the individual participants are risk averse, allowing us to pose the relationship

$$F(t,T) = E_t\{S_T\}.$$

Risk neutral formulation is the basis for most risk management and option pricing theories in commodities markets. The problem with this assumption is that electricity markets are relatively illiquid, with a small number of participants. In light of this we here propose a more general model allowing for the existence of a risk premium in the market. We model the forward price as a function of the spot price, the variance of the spot price, and a random disturbance (z^F):

$$F(t,T) = \Phi(E_t(S_T), \mathrm{var}_t(S_T), z^F)$$

The exact structure of the forward risk premium is likely to vary from market to market.

4.1 IS ELECTRICITY REALLY NON-STORABLE?

A common criticism of the arguments presented above is that electricity really is a storable commodity. After all, the oil, gas and water, which enter into the production of electricity, can all be stored. Therefore a strategy similar to cash and carry arbitrage can be carried out by storing the fuel in times of low electricity prices, and using it to generate electricity when prices are higher. To address these arguments we separate the case of hydro-electric power from the others for reasons that will soon become apparent.

4.1.1 Storage Strategies in Oil and Gas Plants

Consider the owner of a gas fired generator, who purchases the fuel for the plant on the gas spot market, and sells the output on the electricity spot market. We assume that the generator is unable to exercise market power: that is, he takes fuel and electricity prices to be exogenous variables. In our setup, there are two time periods, a current time t and a future time T. For these periods we define the following variables:

S_t^e	The spot price of electricity \$/MW.
S_t^g	The spot price of gas (\$/Btu).
$MC = aS_t^g$	The marginal cost of production for the plant (\$/MW).
q	The capacity of the plant (MW).
$F^e(t,T)$	The forward price of electricity at time t for delivery at time T.
$F^g(t,T)$	The forward price of gas at time t for delivery at time T.
U	The cost of storing gas from t until T (\$/Btu).

Table 4-1

The storage argument will be some variation of the following. Consider the following set of initial prices:

$$aS_t^g > S_t^e$$
$$F^e(t,T) > aF^g(t,T)$$
$$F^g(t,T) > S_t^g + U$$

The first condition states that it is uneconomical to run the plant in the first period based on the gas price. The second condition states that' based on the forward price of gas and electricity, it will be economical to run the plant in the second period. Now consider the case of a generator with no gas storage ability. Its cash flows for the two periods are given in the table below.

	t	T
Plant idle in period t	0	0
Buy a*q gas forwards, for physical delivery	0	$-aqF^g(t,T)$
Sell q electricity forwards, for physical delivery	0	$qF^e(t,T)$
Produce electricity for delivery against forward	0	0
Total cash flow	0	$q\left(F^e(t,T) - aF^g(t-T)\right)$

Table 4-2: Cash flow with no storage option

Next consider the cash flow from the same generator with the option to store gas.

	t	T
Plant idle in period t	0	0
Buy a*q Btu of gas on the spot at t.	$-aqS_t^g$	0
Store gas	$-aqU$	0
Sell q electricity forwards, for physical delivery	0	$qF^e(t,T)$
Produce electricity for delivery against forward	0	0
Total cash flow	$-aq(S_t^g+U)$	$qF^e(t,T)$

Table 4-3: Cash flow with storage option

Assuming a zero discount rate, we have arrived at the following total cash flows:

No Storage	With Storage
$q\left(F^e(t,T)-aF^g(t-T)\right)$	$q\left(F^e(t,T)-aS_t^g\right)$

Table 4-4: Cash flow comparison

Under the third condition of our price levels, the revenue with storage will be higher.

It seems that the owner of the plant has been able to carry out a temporal arbitrage in the electricity market using gas storage. However, if we take a closer look, the strategy is really nothing more than cash and carry arbitrage on the gas market. The condition which must hold in order for the strategy above to be successful,

$$F^g(t,T) > S_t^g + U\ ,$$

is exactly the same constraint which we previously stated could not exist in an arbitrage free gas market. Furthermore, the increase in profit with the storage strategy is exactly the same as the profit from a separate gas arbitrage deal of the same magnitude, and the optimal production strategy for the electricity generator will be the same regardless of whether there is fuel storage capability or not. Assuming that the fuel and electricity spot markets evolve at the same rate, the owner of the plant will always have the option of purchasing additional fuel, or reselling unused fuel on the spot market. There are temporal aspects to operation of these generators, related to minimum and maximum run times and maximum ramp rates. These lead to the so-called unit commitment problem in the production decision problem [7]. It is not possible, however, to exploit or circumvent these constraints through fuel storage.

4.1.2 Storage Strategies in Hydroelectric Dams

Hydroelectric plants are different from gas or oil fired in that their fuel is not a traded commodity. The water flowing into the reservoir does not have an explicit cost, but using the water does represent an opportunity cost to the operator since there is only a limited supply. The operation of a hydroelectric plant is therefore naturally a temporal resource allocation problem under uncertainty. Solving this problem requires advanced dynamic programming techniques.

Some hydro plants are equipped with the ability to act as loads and pump water back up into the reservoir. This setup is known as a pump storage device. It gives the operator the ability to purchase power when the price is low, and produce when the price is high. This process is considered by many to be the equivalent of storing electricity. The inefficiency, or loss of power, in the pump cycle is the equivalent of storage cost. While it is true that the opportunity to capture the difference between high and low price swings adds a temporal component to the electricity production, it is important to differentiate between pump storage and the pure storage of electricity. To illustrate this difference, consider the following example. Note that we use a number of unrealistic simplifications. The purpose is to illustrate the qualitative difference between pump storage and pure storage, rather than provide quantitative results.

We start out with a hydroelectric plant with a maximum output rating of 10MW. The water level in the reservoir is sufficient to produce 10MWh of energy (we will call this quantity 10p.u.). The time scope we are interested in is the next three hours. There will be no new inflow of water during these three hours, and we are not concerned with the amount of water left in the reservoir after this period. Furthermore we take the electricity spot price over these three hours to be known deterministic values: \$10, \$20, and \$30 respectively. Each scenario has a different storage option.

In scenario I, the operator has at his disposal a pumping device which can replace water in the reservoir. The speed of the pump is such that it can replace enough water in one hour to allow the generator to run at full speed for one hour. Furthermore we assume the pump is lossless, that is it will require 10MWh of electricity to run the pump for one hour, exactly equal to the amount which can be generated by this quantity of water. The electricity used in running the pump can be purchased from the spot market. The optimal strategy for operating the plant, and the associated cash flows in each hour, are illustrated in the table below.

Hour	h1	h2	h3	Total
Activity	run pump	run generator	run generator	
Cash Flow	-\$100	\$200	\$300	\$400
Reservoir level at end of hour.	20p.u.	10p.u.	0p.u.	

Table 4-5 Pump Storage Operation

In scenario II we assume the owner has access to a hypothetical pure storage device, in this case a giant battery which can store up to 10MWh of energy. The battery can be charged and discharged at a rate of 10MW. The optimal strategy for running the same hydro plant with this storage option is given below.

Hour	h1	h2	h3	Total
Activity	charge battery	no activity	run generator, discharge battery	
Cash Flow	-$100	$0	$600	$500
Reservoir level at end of hour.	10p.u.	10p.u.	0p.u	
Battery level at end of hour	10MWh	10MWh	0MWh	

Table 4-6: Operation of Perfect Storage Device

We see that the owner of the plant is able to make more money with a pure storage device than with pump storage. This is directly due to the capacity constraint of the generator. No matter how much water is stored in the reservoir, the turbine can only produce power at a rate of 10MW. The pure storage device, however, is able to discharge in parallel with the turbine, thus capturing a larger share of the high price hour. This effect is predominant in electricity markets. Prices are extremely sensitive to the ratio of instantaneous demand and the instantaneous total generation capacity. Therefore, while hydro storage introduces a new inter-temporal component to the electricity price process, it does not add additional capacity, and therefore does not offer the pure arbitrage opportunities available in storable commodities.

4.2 ARBITRAGE AND THE RELATIONSHIP BETWEEN PHYSICAL AND FINANCIAL CONTRACTS FOR ELECTRICITY

One measure of the usefulness of financial markets is the ability of market participants to use financial contracts to hedge physical obligations. In the context of arbitrage, we can use a similar measure. Can a contract for physical delivery of power be perfectly replicated using financial futures or derivatives contracts? Generally this type of replication is a simple exercise, but the special structure of the electricity forward contracts, specifically the difference in the length of the delivery period between the spot and forward markets, makes the replication process somewhat complex.

4.2.1 Application of Arbitrage Pricing Theory in the Relative Pricing of Physical Forward and Financial Futures Contracts on Electricity

We address the relative prices of physical forward and financial futures contracts with ex-post settling, in the framework of arbitrage pricing as defined in the previous section. We allow for the contract to be traded on different exchanges, but assume that there is reasonable price transparency and liquidity in the market. The validity of these assumptions is addressed at the end of the section.

Recall that the notation for the price of a physical forward contract, signed at time t for delivery at time T, is denoted by G(t,T). The equivalent notation for a financial futures contract is F(t,T). We now consider possible relative price levels of the physical and financial markets, and test their consistency with the absence of arbitrage assumption.

First consider the event where, at a time t, we observe a set of contracts for delivery month T satisfying the relationship

$$F(t,T) > G(t,T)$$

A trader can then implement the following strategy.

At time t:

1. Purchase q MW of physical forward contracts.
2. Sell q MW of financial futures contracts.

At time T:

1. For each hour in the delivery period, submit a sell bid of q MW of power into the day-ahead spot market at zero price. The power needed to deliver from the spot market is received from the physical forward contract.

The cash flow from this strategy in each time period is shown in the table. Note that all cash flows from forward contracts are realized at the end of the contract.

	t	T
buy physical	0	$NqF(t,T)$
sell financial	0	$NqF(t,T)-\sum_{i=1}^{N} qS_i$
sell spot	0	$\sum_{i=1}^{N} qS_i$
Total	0	$Nq(F(t,T)-G(t,T))>0$

Table 4-7: Arbitrage in Electricity Forward Markets

This strategy provides a guaranteed profit with zero investment, and therefore it is an arbitrage opportunity which cannot be sustained.

Now consider the case

$$F(t,T)<G(t,T).$$

The trader adopts the following strategy.

At time t:

1. Purchase q MW of financial futures contracts.
2. Sell q MW of physical forward contracts.

At time T:

1. For every hour in the delivery period, submit a buy bid for q MW to the day-ahead spot market at the market maximum price (we later discuss what happens if the spot market fails to clear). The electricity purchased in the spot market is used to deliver against the obligation from the physical forward contract.

The cash flows in each time period are given by:

	t	T
sell physical	0	$-NqF(t,T)$
buy financial	0	$\sum_{i=1}^{N} qS_i - NqF(t,T)$
sell spot	0	$\sum_{i=1}^{N} qS_i$
Total	0	$Nq(G(t,T) - F(t,T)) > 0$

Table 4-8: Arbitrage in Electricity Forward Markets

This strategy provides a guaranteed profit with zero investment, and therefore it is an arbitrage opportunity which cannot be sustained.

The strategies presented above show that in a market free of arbitrage opportunities, the price of a financial forward cannot deviate from the price of a physical forward, in either a positive or negative direction. This condition must hold true not just at maturity, but during the entire lifetime of the contracts. We thus arrive at the first constraint for electricity derivatives in an arbitrage free marketplace:

$$G(t,T) = F(t,T) \ \forall\, t,T$$

Based on this constraint we will from now on use physical and financial forward contracts interchangeably.

Chapter 5

Building a Price Model for Electricity Markets

In the previous chapters we have outlined the dynamic constraints which restrict the evolution of market price, based on arbitrage theory, replication strategies, market agents' utility functions, and physical constraints on production and storage. In this chapter we attempt to build a price model for the electricity industry. Based on the lessons learned from other industries, we will use the following criteria in evaluating the effectiveness of the model:

1. The model must reflect reality in a dynamic and probabilistic setting. This includes properly reflecting the probability distributions and covariance matrices of future electricity prices.
2. The model must have a form conducive to solving the stochastic optimization problems facing the electricity industry, including valuating generation assets, hedging long term contracts, and managing locational risk. This involves addressing the unique properties of how electricity is produced, transmitted and consumed. Specifically it must address the inter-temporal relationships of prices for a non-storable commodity, the locational price differences in complex transmission networks, and the inter-commodity dynamics of electricity, fuel and emissions rights markets, which influence the cost of production for generators. Designing the model so that it can easily be expanded in the temporal, special and inter-commodity domain is therefore crucial to its performance.
3. The complexity and form of the model must be limited so that it can be used as an input to standard optimization algorithms such as dynamic programming. This constraint is often in direct conflict with (1) and (2), which tend to add complexity in order to better reflect reality.
4. The model must be structured so that it can be consistently calibrated from available data. If a trader or investor can not trust that the

parameters of the model reflect his current knowledge of the market he will have little confidence in the decision rule generated by any optimization technique. This requirement is particularly crucial in new markets such as electricity, where price history is limited, and where market structure as well as regulatory guidelines are constantly evolving.

5.1 STRUCTURE OF MODEL

Several times in this book we have touched on the question of how best to deal with the dimensionality of price dynamics in electricity markets. This includes defining the dynamic and stochastic relationship between different time periods, different locations in the network, different markets (spot, forward, derivatives), and different commodities (electricity, oil, gas). We found that on the basis of arbitrage pricing theory, we cannot impose constraints on the temporal relationship of prices. This is a direct result of the non-storability of electricity. Similarly, it is not possible to create pure arbitrage strategies between electricity and fuel markets, or between the electricity prices at different locations in the network, except in degenerate cases. From a strictly theoretical point of view, we could therefore conclude that each spot, forward and derivative contract, for each delivery period, at every location in the network, should be modeled as a separate state of a dynamic process. This modeling approach, however, is highly impractical. The dimension of the state space would be so high that we would be unable to solve even the simplest optimization problems, not to speak of defining consistent methods of estimating model parameters. The focus in this section of the book therefore is on finding a reasonable compromise: a model which preserves the unique characteristics of electricity production, consumption and transmission, while limiting complexity. To achieve this goal, we have identified a set of fundamental drivers. These are external processes, physical as well as financial, which have significant impact on the dynamics of electricity prices, and whose effects on the supply of, and demand for, electricity are reasonably well understood.

In characterizing the interaction between the spot, forward and derivatives markets, we have taken the approach of modeling forward and derivative prices as outputs of the basic spot market process. We recognize that this approach, in contrast to equity or storable commodities markets, is not backed by a theoretical arbitrage argument. We therefore need to allow more leeway in terms of the presence of risk premium, and market specific risk, than is common in other markets.

Finally we recognize that because of the unique temporal constraints on supply and demand in electricity markets, the problem lends itself naturally to time scale separation arguments. Specifically we address three time scales:

- Fast deviations, resulting mainly from temperature driven load fluctuations, and strategic bidding or unit outages on the supply side. These processes are mean reverting with fast time constants.
- Medium term deviations, including fuel price changes and aggregate rain fall (hydro reservoir levels), which revert slower or not at all.
- Long term deviations, mainly demand growth and new generation investment. This time scale differentiates itself mainly because it includes price feedback with substantial delays. This leads to dynamic behavior not generally addressed in financial models.

Figure 5-1 illustrates the full blown flow of physical and financial signals in the model. As we proceed to model the dynamics of the system, we identify how the interplay between the components can be broken down based on the time scale at which the interaction occurs. This allows us to derive simpler versions of the models for use in specific applications. For example, a user interested in developing day-ahead bidding strategies or short term price hedging strategies need not be concerned with load growth and investment dynamics. It is crucial, however, to start with the full scale version of the model in order to understand the inter-temporal dynamics, and thus have a solid basis for any separation or model reduction arguments.

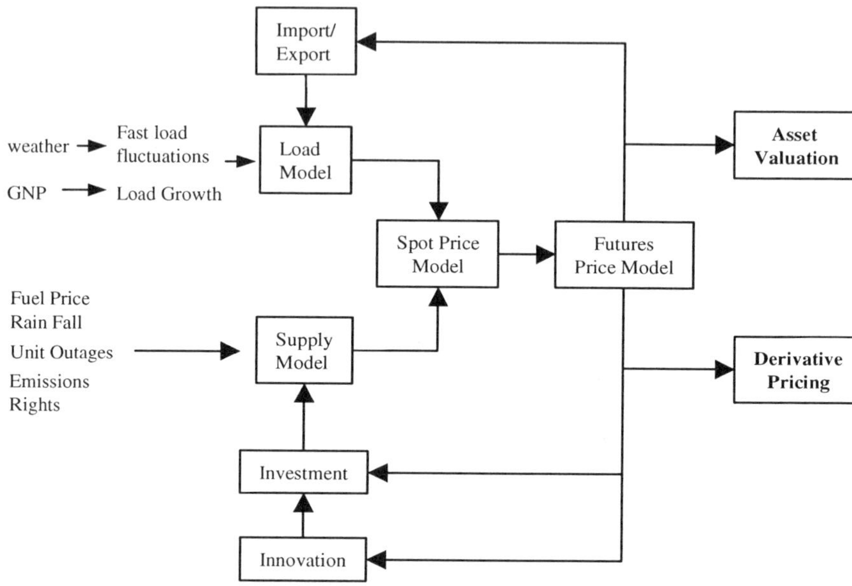

Figure 5-1: Dynamic Relationships in Electricity Markets

5.2 MODELING APPROACHES

In this section we outline a number of approaches which have been used to model price processes in electricity markets. Each approach has its own distinct set of advantages and disadvantages. All of them represent valid approaches to understanding the evolution of prices in the market, and the choice of model is inherently related to the specific purpose of the user.

5.2.1 Quantitative (Statistical) Modeling of Electricity Prices [5],[8],[8],[10]

Objectives: To characterize the stochastic properties of commodity prices over time; specifically, to attempt to derive the variance and covariance of commodities prices. With this information the user is then able to price a broad category of financial derivatives, as well as perform basic risk management functions.

Characteristics: The models used in quantitative modeling are usually generic in nature. The user attempts to find the lowest order model possible to describe accurately the stochastic properties of the commodity.

Advantages: Since the model is generic, the user does not require an in-depth understanding of the economic or physical relationships involving the production and trading of the commodity. Calibration schemes are also standardized and can be duplicated across multiple commodities.

Disadvantages: This category of models is calibrated using historical spot and forward market data and, when available, implied volatilities from historical models. It requires the availability of a significant amount of price history data. In the case of electricity, changes in the regulatory environment have made historic prices invalid for calibration purposes, leaving the user with an inadequate set of training data for the models.

5.2.2 Production (Cost) Based Modeling of Electricity Prices

Objectives: To model future electricity prices based on detailed models of the cost structure of individual products. This information is used to create a cost-based supply curve. Combined with estimates of future demand, this can be used to generate price estimates.

Advantages: Marginal cost information is generally available for all producers in a region. The creation of a supply function is therefore a relatively straightforward exercise. Furthermore, the cost can be linked to underlying fuel prices by using heat rate estimates on the unit. This allows the user to model the interaction of fuel and electricity prices.

Disadvantages: Cost based modeling ignores the strategic bidding practices of market participants. The effect of market power is likely to raise prices above cost-based levels. The cost-based models can therefore rarely be calibrated to correspond to actual observed prices in the market.

5.2.3 Economic Equilibrium Models of Electricity Prices [11],[12],[13],[14]

Characteristics: As a means of incorporating strategic bidding into cost-based models, theories such as Cournot pricing are applied to the generation stack. At a given load level one can then solve for an equilibrium markup of bids above cost based levels. This markup will generally increase as a function of market concentration.

Advantages: By applying game theory type models it is possible to explain why prices rise above cost-based levels. This approach is useful in predicting expected price levels in markets with no price history, but with known supply costs and market concentration.

Disadvantages: These models produce equilibrium price levels. However, electricity markets are constantly evolving, driven by stochastic supply and demand, and therefore never settle to equilibrium levels. In applications such as risk management, understanding the dynamic behavior of prices is crucial. In this case economic equilibrium models offer little insight.

5.2.4 Agent-based Modeling of Electricity Prices [15],[16]

Characteristics: Agent-based models attempt to capture the strategic behavior of investors (agents) in the marketplace. To approximate the dynamics of the market, participants are separated into groups, each with their own objective function. Based on the objective function and observation of current price levels, a decision rule is defined for each group. These rules can be highly nonlinear in nature. Finally the system is simulated under various inputs.

Advantages: In contrast to cost-based and equilibrium models, agent-based models address the effect of market power both on the overall price markup, and on the inter-temporal dynamics of price. The variety of dynamic behavior, which can be captured with a relatively small number of strategies, is impressive. The approach, for example, allows the user to study the impact of factors such as collusion on the overall system price.

Disadvantages: While agent-based modeling is useful for studying the qualitative behavior of markets, it also makes it much more challenging to get relevant quantitative results. To do so, one would need a consistent method of calibrating the parameters of the decision processes based on historical data. This seems like an overwhelmingly difficult task.

5.2.5 Experimental Modeling of Electricity Prices [17]

Characteristics: In the experimental modeling approach, a group of people are gathered and assigned assets and obligations in the marketplace. They then simulate the behavior of the market by submitting bids, which are used to clear the market.

Advantages: The organizer of the experiment has full control over the parameters and can change factors such as market concentration or number of participants in order to observe the effects on the spot price.

Disadvantages: Experimental modeling is extremely difficult to map into a real marketplace. To get reliable results, one needs to convince actual marketers to participate in the process, and even then it is questionable if they will betray their actual trading strategies.

5.2.6 Fundamental Modeling of Electricity Prices [18],[19],[20]

Objectives: Determining the stochastic properties of commodities prices.

Characteristics: In the fundamental modeling approach, price dynamics are described by modeling the impact of important physical and economic factors on the commodity price. The model seeks to capture basic physical and economic relationships present in the production and trading of the commodity. By explicitly adding these constraints, one can increase the complexity of the model while decreasing the requirements on the available training data.

Advantages: By relaying the dynamics of the commodity price to the fundamental drivers, we gain a new set of training data. If the fundamental inputs are directly observable, we can use historical inputs in order to calibrate the model parameters. In the case of electricity this can be a crucial difference. Currently there are only between two and tree years of relevant electricity price history available (depending on the location). However, if we choose temperature (a major determinant of electricity demand) as a fundamental driver, we have decades worth of historical measurements available.

Disadvantages: In creating the fundamental model we make specific assumptions about economic relationships in the marketplace. The price projections generated by the models are therefore very sensitive to violations of these assumptions. Thus there exists a significant modeling risk in the application of the fundamental approach.

Chapter 6

A Bid-based Stochastic Model for Electricity Prices

In this chapter we develop a bid-based stochastic model (BSM) of the evolution of prices on electricity spot markets [19]. We assume that the spot market operates as a double auction, with a single hourly market clearing price (MCP) at the intersection of the aggregate supply and demand bid curves.. The model can be modified to account for variations in the auction procedure. We design the model to be applicable to hedging, speculation or investment decisions in electricity markets. As such, it focuses on quantifying the uncertainty of future price movements. We have used a fundamental modeling approach, where the fundamental drivers are load and supply shifts. The model captures the most critical characteristics of demand (load) and supply, as outlined below, in the electricity market.

6.1 LOAD CHARACTERISTICS

1. **Load Elasticity**: We assume electricity demand to be completely inelastic (i.e. independent of the market clearing price). This may appear to be a strong assumption, but in the current state of deregulation, few end users actually observe real time price movements.
2. **Seasonality**: Seasonality is a major driver for electricity demand. We observe seasonality over the daily, weekly, and yearly cycles.
3. **Mean reversion**: One can observe temporary spikes in electricity demand, often induced by extreme weather conditions. However, these spikes are not sustainable and demand reverts back to normal levels within a few days.

4. **Stochastic growth**: Growth in electricity demand is driven in part by trends in the overall economy. The growth is therefore hard to predict over longer time horizons, and must be considered stochastic.

6.2 SUPPLY CHARACTERISTICS

1. **Supply Elasticity**: In contrast to load, electricity producers are price responsive. The supply characteristic is mainly a function of generation technology, as operating cost can vary significantly with the type of generator used. Market power and strategic bidding also have an impact on the shape of the supply bid function.
2. **Stochastic Availability of Generation**: Due to unexpected equipment failure or because of planned maintenance, generators are taken offline from time to time. The effect of such sudden jumps in the availability of supply on the market clearing price can be significant.
3. **Uncertain fuel cost**: Changes in the price of fuels such as oil and gas will affect the way generators bid into the market.
4. **Unit Commitment**: Nonlinear characteristics in the generator cost functions, such as startup costs and minimum run times, result in intra-day supply bid curve shifts.
5. **Import/Export**: Producers and consumers bidding into the market from outside its geographic limits can cause significant price shifts.
6. **Inter-Market**: Prices in related markets such as the markets for capacity and ancillary services, represent opportunity costs for power suppliers. Hence there is a strong interaction between prices in these markets and the price in the energy market.

6.3 PRICE AS A FUNCTION OF LOAD AND SUPPLY

In our model we characterize spot price as a function of two variables: L representing load shifts, and b representing supply shifts. These variables can be interpreted as follows.

Load: We assume load bids are inelastic. Therefore L_k is selected to represent the market clearing volume of the exchange for hour k.

Supply: In contrast to load, supply bids have significant price elasticity. The elasticity (or the inverse of the slope of the supply curve) varies significantly with the clearing quantity. In general, supply will be highly elastic at low demand levels, and gradually become more inelastic as demand increases.

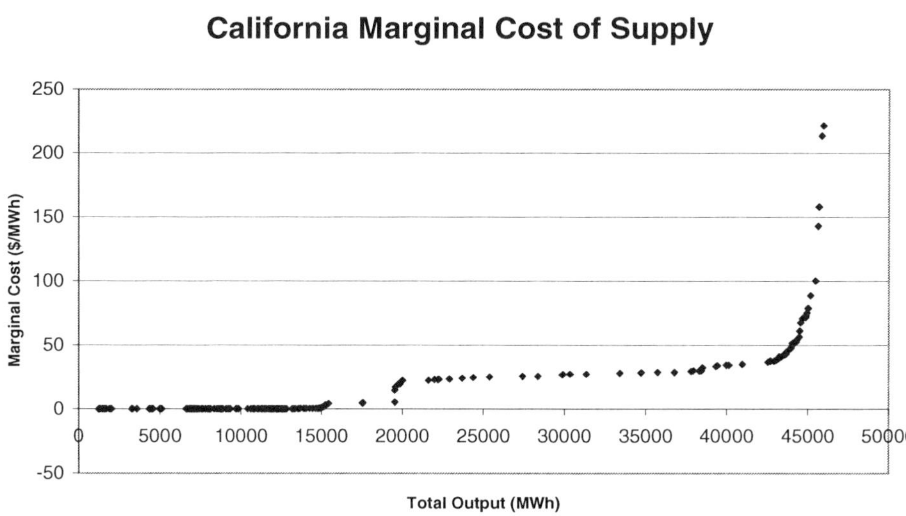

Figure 6-1 Marginal Cost Stack for California

We can explain this characteristic of electricity supply in two ways. First we examine the cost structure of the underlying generators. Figure 6-1 shows the 'stack' for California (data from 1998 assuming corresponding fuel prices), created by ordering the generators from lowest to highest marginal cost. As seen in the figure the cost function is relatively flat for low demand levels, when the load is served mainly by hydro and nuclear plants. In the medium range we see a slight cost increase as efficient fossil plants are utilized. In the high demand range, inefficient peaking plants are dispatched, and the operating cost escalates significantly.

Another approach is to view the supply bids from a game-theoretic perspective. At low demand levels there is a high ratio of available generation capacity to electricity demand. Hence the market will be competitive and highly price responsive. As load approaches the total installed capacity of the market, the few non-committed generators have a high degree of market power, and can withhold their capacity to push prices upward. An in-depth analysis of market power and strategic bidding in power markets can be found in [15] and [16].

The next figure shows a plot of the supply curves submitted to the California Power Exchange during a 24-hour period in 1998.

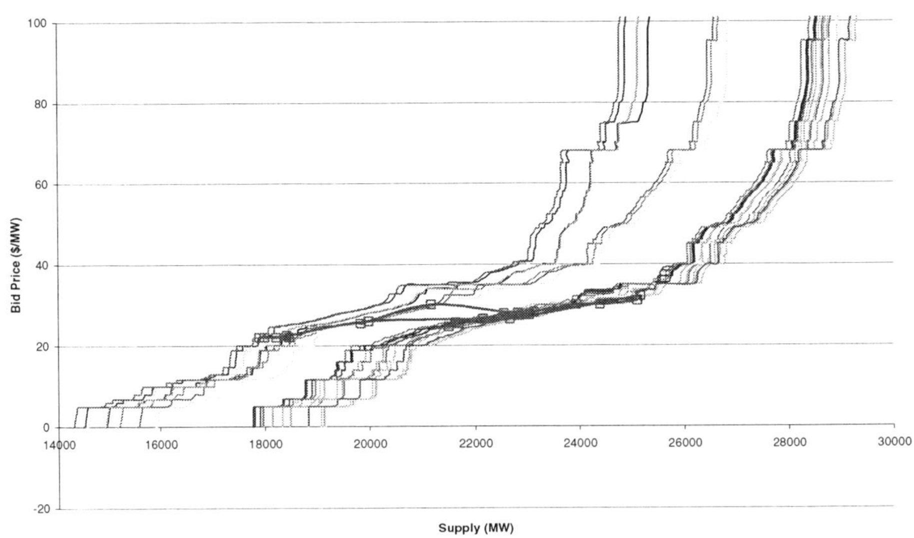

Figure 6-2 Aggregate supply bid curves into the CalPX.

When comparing the cumulative bid curves submitted at different hours we find that the basic shape of the bid curve is preserved over time. This allows us to reduce the complexity of the supply model. We fix the shape of the bid curve and model its temporal shifts as a stochastic process. Specifically we choose an exponential function to approximate the shape. Price in hour k can then be written as

$$S_k = e^{aq_k + b_k}$$

where a is a fixed parameter characterizing the slope of the bid curve, q_k is the market clearing quantity in hour k, and b_k denotes the position (or shift) of the curve. Next we add the constraint that demand bids are inelastic. The market clearing quantity q_k must then always be equal to the system load L_k. We can now write the market clearing price in terms of our two fundamental drivers, load and supply:

$$S_k = e^{aL_k + b_k}.$$

This approach reduces the complexity of the problem by constraining the number of free variables on the supply side.

The next step is to postulate stochastic models for the evolution of the fundamental drivers. In order to keep track of the variables and parameters evolving at different time scales we use the following notation:

Subscript	Meaning	Superscript	Meaning
d	evolves at daily rate	L	Belongs to load process
m	evolves at monthly rate	b	Belongs to supply shift process
none	constants	δL	Applies to load mean process δ^L
none	constants	δb	Applies to supply mean process δ^b

Table 6-1: Notation for bid based price model

The following section will outline the models used and the reason for choosing that specific form. In later sections we present step-by-step descriptions on how model parameters were calibrated based on historical market data.

6.4 STOCHASTIC LOAD MODEL

We listed the four characteristics of electricity demand which we want to capture in the model: lack of price elasticity, seasonality, mean reversion and stochastic growth. The elasticity assumption is already implicit in our formula for market clearing price P_k. The challenge is to incorporate the remaining criteria without making the model too complex for calibration and simulation purposes.

6.4.1 Modeling Demand Seasonality

The three types of seasonality in electricity demand are reflected in daily, weekly and yearly patterns. Figure 6-4 shows the demand in New England for a sample week in May, starting with Monday.

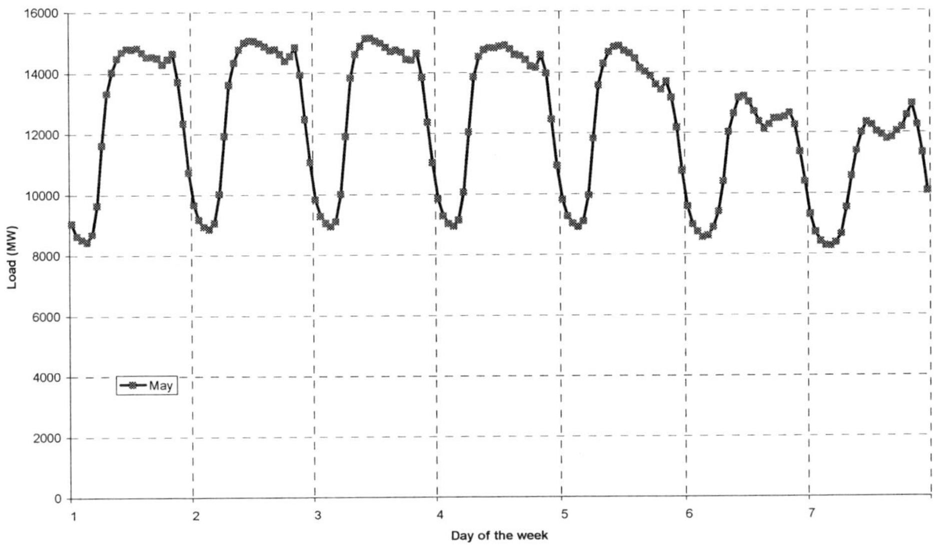

Figure 6-4: Hourly loads for a week in May, New England

We see that there is a regularly recurring pattern within the weekdays (daily seasonality) and that the weekend consumption pattern is significantly different (weekly seasonality). From now on we will simplify our task by eliminating the weekends and modeling only the weekday loads. This allows us to ignore the weekly seasonality. This simplification is taken directly from the forward markets, which typically trade weekdays and weekends as separate contracts.

Addressing the daily seasonality is more challenging. We have chosen to denote the daily load as a [24×1] vector $\mathbf{L}_d$, where each component represents an hourly load. This vector is defined as the sum of a deterministic and a stochastic component:

$$\mathbf{L}_d = \boldsymbol{\mu}_m^L + \mathbf{r}_d^L .$$

The deterministic component μ_m^L is a [24×1] vector that represents the typical or average monthly load pattern for the day. A separate vector μ_m^L is calibrated for each calendar month, in order to capture the yearly seasonality.

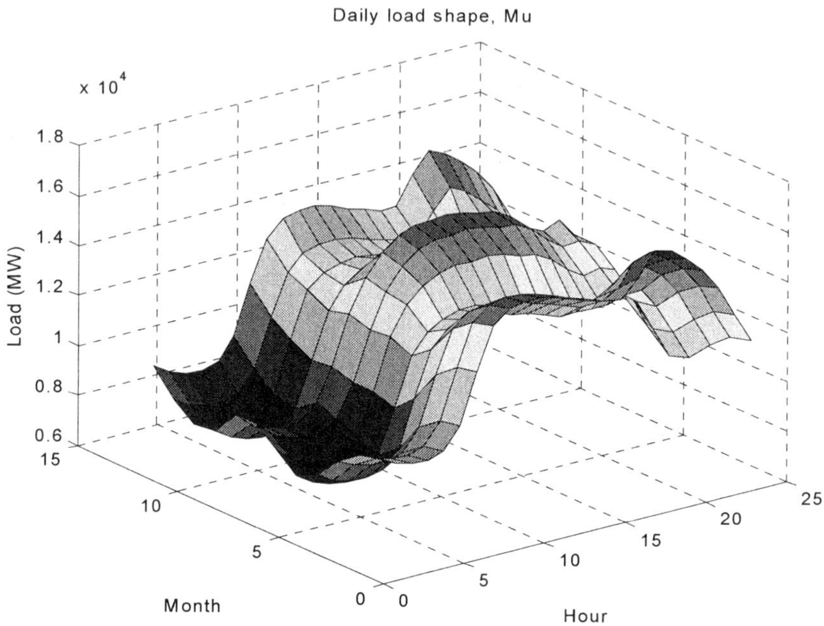

Fig. 6-5: Average monthly patterns of daily load, $\mu_m{}^L$, New England

6.4.2 Modeling Load Uncertainty

The stochastic component **r** of the daily load pattern is needed to explain any deviation in the actual observed load from the pattern given by μ_m. In order to achieve this, the vector **r** has to contain 24 random variables. However, in observing actual load patterns, one finds that there is a strong correlation between deviations in consecutive hours. Intuitively, one could argue that if unusually hot weather causes demand to increase in hour 14, it is very likely to cause higher demand in hours 15 and 16 as well. To capture this mathematically, we applied Principal Component Analysis (PCA) to the data; see [21],[22] and [23]. Principal Component Analysis is a method that enables us to describe a set of observations of n variables, which would normally require n dimensional representation, with a reduced set of j variables, $j \leq n$.

In other words, PCA addresses the issue of how to characterize a probabilistic space of n dimensions using a reduced set of j basis functions.

Although some information will be lost in this process, PCA enables us to minimize this information loss by choosing the new basis intelligently. PCA generates a set of new basis vectors, called Principal Components (PC). All PCs are orthogonal to each other, forming a new orthogonal basis. The theory supporting principal components and their derivation is provided as an Appendix.

The output of the PCA algorithm is a set of principal components $\mathbf{v}^{Li}$ and associated weights w^i, chosen so as to minimize the mean squared error between the observed historical load vectors, and the projection onto the new basis is given by

$$\mathbf{L}_d = \boldsymbol{\mu}_m^L + \sum_{i=1}^{j} w_d^{Li} \mathbf{v}_m^{Li} .$$

In this report we will use a single principal component, a monthly [24×1] vector $\mathbf{v}_L{}^m$, to describe load behavior, reducing the load equation to

$$\mathbf{L}_d = \boldsymbol{\mu}_m^L + w_d^L \mathbf{v}_m^L ,$$

where μ_m and $\mathbf{v}_m$ are deterministic parameters and w_d is a daily stochastic process.

The choice of the number of principal components used is a tradeoff between accuracy and complexity. For short-term decision making such as day-ahead bidding, a single PC may not provide a rich enough sample space. However, when applying the price process to hedging and valuation decisions over months or years, a small basis prevents the problem from blowing up into computational complexity.

Next we need to address how the stochastic component w_d evolves over time. The model we propose is a two factor mean reverting model:

$$e^L_{d+1} - e^L_d = -\alpha^L e^L_d + \sigma^L_m z^L_d$$
$$\delta^L_{d+1} - \delta^L_d = \kappa^L + \sigma^{L\delta} z^{L\delta}_d ,$$

where

$$e^L_d = w^L_d - \delta^L_d .$$

6.4.3 Mean Reversion

We can interpret the states e_d^L and δ_d^L in terms of the temporal characteristics of load. The state e_d^L models short term deviations in load, such as those caused by sudden heat waves. These events are generally temporary, and load gradually reverts back to normal levels. The process for e_d^L is therefore chosen to be mean reverting. The parameter α determines the speed of reversion. Figure 6-6 illustrates how the short-term spikes in load quickly revert to the long-term mean. For clarity, here the mean is being modeled as a monthly rather than a daily process. This time scale separation between the states is a method which further simplifies the application of the model, and its advantages and disadvantages are described in the calibration section.

6.4.4 Stochastic Growth

As e^L reverts to zero, the weight w^L reverts to δ^L, or the "normal" load level. However, since the market is never at an equilibrium, the normal load level is in itself a stochastic process. The δ^L process characterizes the stochastic growth of load over time. This growth could be positive or negative for any given period of time, and there can be significant uncertainty to the rate of growth, captured in the long-term volatility parameter $\sigma^{L\delta}$. The long-term growth of load in New England is illustrated in Figure 6-7.

The effects of the structure of the stochastic process on the mean and variance of future load is explored in detail in the section on simulation.

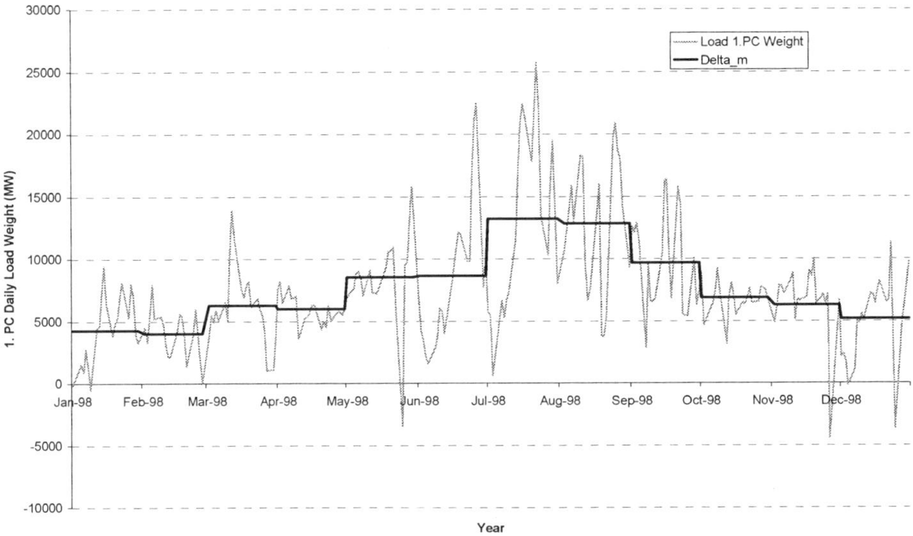

Figure 6-6 Reversion of the load weights w_d^L to the long-term mean δ_m^L, year 1998

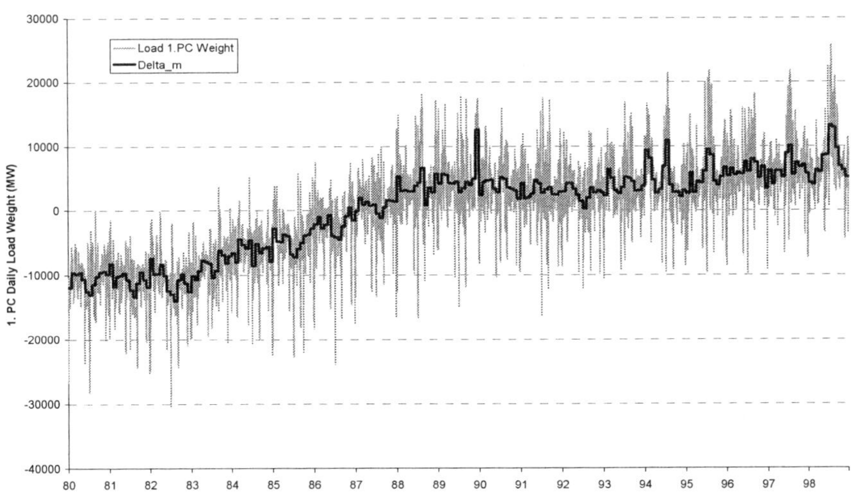

Figure 6-7 Load weights w_d^L, and long-term mean δ_m^L, New England

6.5 STOCHASTIC SUPPLY PROCESS

Recall our underlying price model as a function of load and supply states L_k and b_k:

$$S_k = e^{aL_k + b_k}.$$

This implies that the aggregate supply bid curve is an exponential function of fixed shape (given by a), which shifts over time.

Let us consider the input drivers, which could cause the supply curve to shift:

1. **Fuel price**: An increase in fuel prices would force suppliers to increase their bids into the spot market in order to remain profitable. An increase in the fuel price would therefore be accompanied by a positive shift in b_k.

2. **Unit Outages and Scheduled Maintenance**: The withdrawal of a generation unit from the market, whether through an unexpected failure or a scheduled maintenance, causes a significant shift in the supply bid function. The size and duration of such a shift, as well as the frequency of their occurrence, is technology dependent.

3. **Gaming and Strategic Bidding**: It has been shown that generators with significant market share may increase their profits by unexpectedly removing part of their generation assets from the market, forcing up price and increasing the payoff for the remaining units [36]. Such an event can be characterized by a positive shift in b_k.

4. **Unit Commitment Decisions**: While generators are often modeled as having well behaved quadratic cost functions, in reality there are significant non-standard costs and constraints associated with starting up and shutting down a generator. Translating such constraints into bids will cause generators, even though they may have no market power, to deviate from marginal cost-bidding schemes.

We now attempt to translate the impact of these drivers into a stochastic process for the supply process. As with the load we characterize supply by a [24×1] daily vector $\mathbf{b}_d$ containing hourly supply levels. This daily vector is then decomposed into its deterministic and random components:

$$\mathbf{b}_d = \boldsymbol{\mu}_m^b + \mathbf{r}_d^b$$

6.5.1 Seasonality of Supply

Although less pronounced than the load, the supply process does exhibit seasonality over multiple time scales. The most pronounced are monthly and intra-day seasonality.

1. On a **monthly** time scale, we see the scheduling of maintenance. In a practice that has carried over from the regulated industry, units are regularly scheduled for maintenance during the off-peak seasons (mainly fall and spring), when demand spikes are unlikely. From the modeling perspective this creates a repeating twelve-month pattern of supply bid shifts.

   The fuel markets feeding the generators also experience seasonality on this time scale, mainly due to seasonal demand for oil and gas. Seasonal fuel prices therefore create a second pattern of supply shifts. The aggregate effect of these repeating yearly patterns is captured by the deterministic shifts in the monthly parameter μ_m.

   Regions with significant amounts of hydro generation may exhibit a different type of seasonality, since the level of water in the reservoirs is linked to precipitation patterns, and the melting of snow caps.
2. The second time scale in which seasonality is observed in the supply process is **intra-day**, where we observe repeating 24-hour patterns of supply curve shifts. This type of seasonality is mainly attributed to unit commitment decisions made by the dispatchers. The operator of the unit will estimate a day ahead of time the hours during which it will be profitable to run the unit, based on the startup/shutdown constraint of the

generator. Once this decision is made he may choose not to submit bids for the remaining hours, so as not to risk being scheduled and incurring a substantial startup cost. The result is a repeated pattern of hourly shifts in the aggregate supply bid curve. This behavior is captured by the daily shape of the vector μ_m^b.

6.5.2 Modeling Supply Uncertainty

In modeling the random component of the daily supply vector we again apply principal component analysis, using a first order approximation (one PCA vector):

$$\mathbf{b}_d = \boldsymbol{\mu}_m^b + w_d^b \mathbf{v}_m^b .$$

The process defining the evolution of the weights is similar to that used for the load process:

$$e_{d+1}^b - e_d^b = -\alpha^b e_d^b + \sigma_m^b z_d^b$$
$$\delta_{d+1}^b - \delta_d^b = \kappa^b + \sigma^{b\delta} z_d^{b\delta} ,$$

where

$$e_d^b = w_d^b - \delta_d^b .$$

The mean reverting component e_d^b reflects the transient characteristics of the supply process. This includes short-term fuel price spikes and short-term gaming. These effects are temporary and die out at a rate governed by α. The reversion of supply to its long-term mean is illustrated in Figure 6-8:

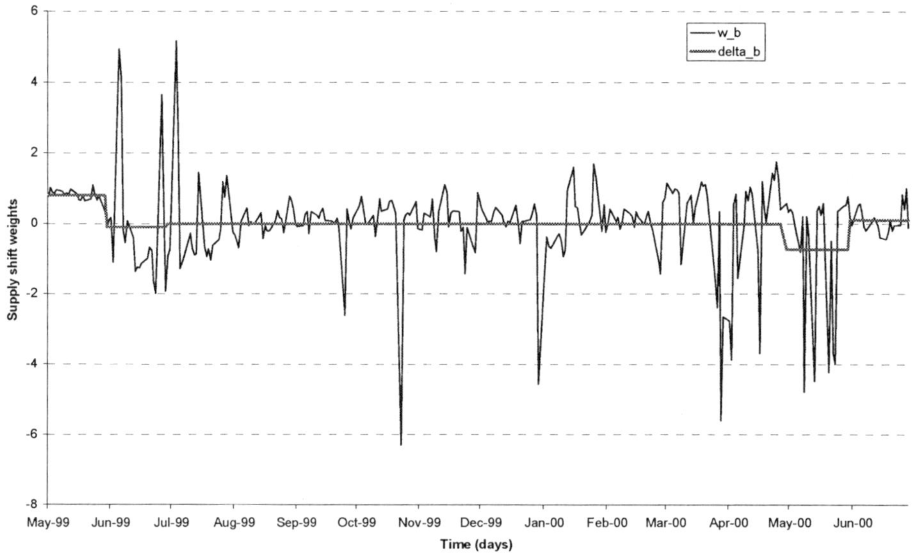

Figure 6-8 Reversion of supply weights w_d^b to long-term mean δ_m^b

The non-reverting component δ_d^b models the long-term availability of generation. This will include any new installed or retired capacity in the market.

6.5.3 Modeling Unit Outages

So far our supply model has included smooth changes in the behavior of the supply bid curve, which can be characterized by an Ito process. However, there exists a set of high impact, low probability events that cannot be approximated through random walk type models. One such event is the unexpected failure of a major generator in the market. There are a number of unknowns associated with this event:

1. The probability of an outage on a given day.
2. The impact of the outage on market price.
3. The duration of the outage.

The answers to all three of these questions generally depend on the type of generation technology.

In our model we address these problems by adding a new factor to the supply process:

$$\mathbf{b}_d = \boldsymbol{\mu}_m^b + w_d^b \mathbf{v}_m^b + \sum_i \pi_d^i \boldsymbol{\psi}_m^i .$$

1. The probability of an outage occurring on a given day is modeled as a random incidence process, specifically a Bernoulli process. The probability of the outage occurring on a given day is independent of all other time intervals. This is denoted by the variable π^i, where π^i = 0 under normal conditions, and π^i = 1 when there is an outage in a plant of technology i.

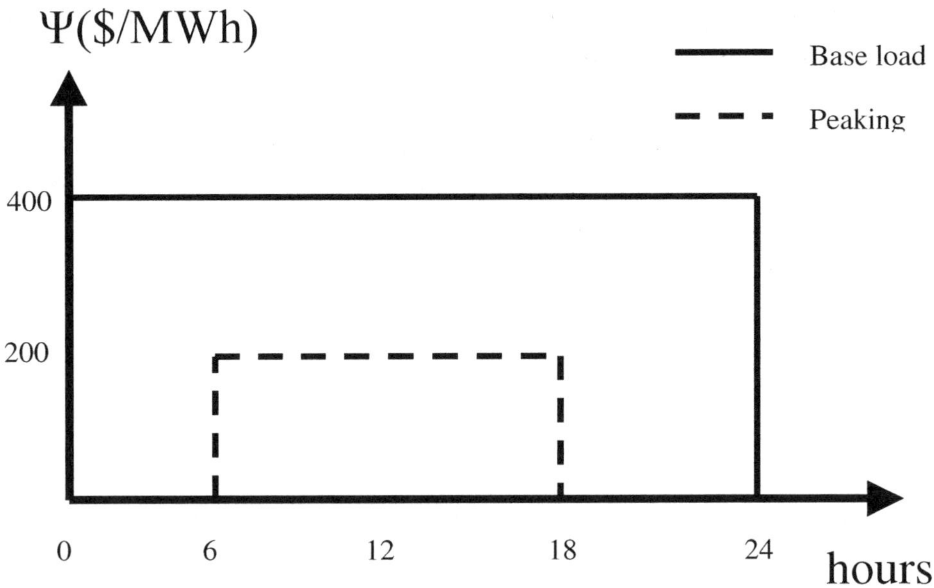

Fig. 6-9 Daily shape ψ for a 400 MW base load plant and a 200 MW peaking plant

2. The impact of the outage on the market clearing price will depend on the capacity of the unit and its characteristic operating schedule. An outage in a plant which is scheduled to deliver at full capacity results in a positive shift in $\mathbf{b}_d$, equal to the capacity of the plant. If, however, the plant was not scheduled to deliver (i.e. bid in above market clearing price) then there is no effect on the price. The probability of a plant being selected to produce in a given hour generally depends on its cost structure, and therefore on its technology. We incorporate this effect by assigning a [24×1] vector ψ_m^i to each technology i. The vector denotes the capacity of the unit as well as the likelihood of the unit being scheduled in a given hour. Figure 6-9 denotes the daily shape of ψ for two types of generation technologies, a 400 MW base load plant and a 200 MW peaking plant.
3. The outage duration is modeled as a deterministic minimum outage time plus a stochastic Bernoulli component. By combining this

process with the random arrival time of the outage, we can characterize the process for the state π^i_d as a Markov chain, as illustrated in Figure 6-10. Here the numbers next to the arrows designate the probability of a state transition for a given day. The probability of going from normal operation to an outage for each day is given by λ_{out}. The probability of the unit returning on-line after the minimum outage period is given by λ_{in}. For the case shown the minimum outage time is four days.

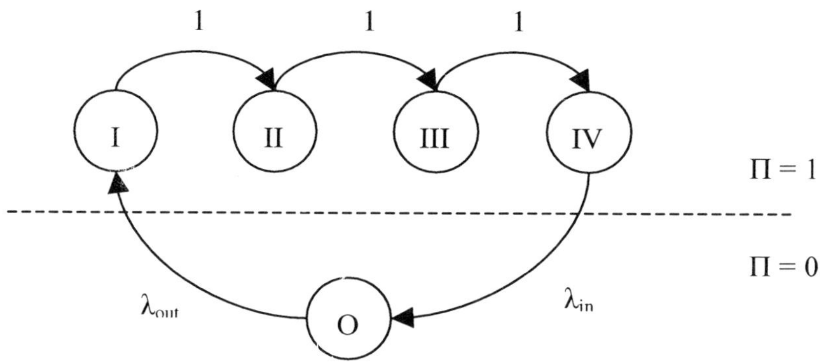

Figure 6-10: Modeling of outage duration for the state π^i_d as a Markov chain

6.5.4 Modeling Scheduled Maintenance

Scheduled maintenance can be modeled in the same manner as unit outages. The only difference is that the π_d becomes a deterministic rather than stochastic state variable.

6.6 SUMMARY OF THE BID-BASED STOCHASTIC PRICE MODEL

The following is a compact summary of the mathematical model underlying the Bid-based Stochastic Model.

Spot Price Model:

Hourly price: $S_h = e^{aL_h + b_h}$

Daily 24-hour vector of prices: $S_d = e^{a\mathbf{L}_d + \mathbf{b}_d}$

Load Model:

$$\mathbf{L}_d = \boldsymbol{\mu}_m^L + w_d^L \mathbf{v}_m^L,$$

$$e_{d+1}^L - e_d^L = -\alpha^L e_d^L + \sigma_m^L z_d^L$$

$$\delta_{d+1}^L - \delta_d^L = \kappa^L + \sigma^{L\delta} z_d^{L\delta},$$

where,

$$e_d^L = w_d^L - \delta_d^L.$$

Supply Model:

$$\mathbf{b}_d = \boldsymbol{\mu}_m^b + w_d^b \mathbf{v}_m^b + \sum_i \pi_d^i \psi_m^i.$$

$$e_{d+1}^b - e_d^b = -\alpha^b e_d^b + \sigma_m^b z_d^b$$

$$\delta_{d+1}^b - \delta_d^b = \kappa^b + \sigma^{b\delta} z_d^{b\delta},$$

where,

$$e_d^b = w_d^b - \delta_d^b.$$

and π_d is Markov process with parameters λ_{out} and λ_{in} as described in the previous section.

6.7 CALIBRATION OF THE BID-BASED STOCHASTIC MODEL

In this section we describe the process of calibrating the model based on publicly available market data. For this purpose we used historical hourly load and price data for the New England region, available on the New England ISO home page. The available load data spanned a 20 year period from 1980 to 2000, while the price data was limited to the fourteen month period since the inception of the market: May 1999 to June 2000. Based on this data, the parameters of the bid based model presented in this chapter were estimated. The only deviation from the model was that the unit outage component was omitted due to a lack of available historical outage data. The stepwise process of calibrating the model is outlined in the figure 6-11.

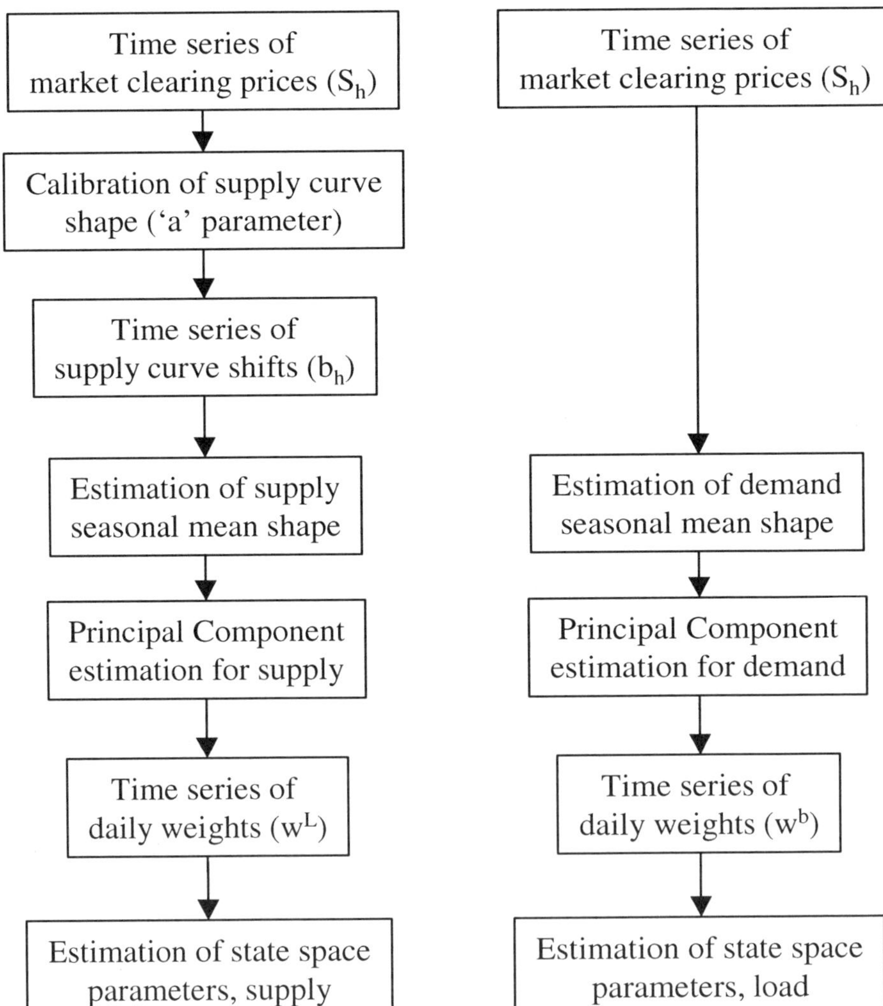

Figure 6-11 Flow chart of the calibration process

6.7.1 Generating a Time History of Supply States

As we have discussed previously, the hourly price dynamics in our model are governed by a combination of supply and demand states, L and b respectively, characterized by the relationship

$$S_h = e^{aL_h + b_h}.$$

The parameters describing the load state dynamics can estimated directly from the time history of the market demand. The supply state b_h, on the other hand, represents the hourly position of the aggregate supply bid curve, and as such is not directly observable. To overcome this problem we use the historical load and price data to calculate the implied time history of the supply state, given by the equation

$$b_h = \ln(S_h) - aL_h.$$

This approach requires us first to estimate the supply curve shape parameter 'a'. There are several possible approaches to determining this shape. In markets where the supply bids are published (as is the case currently for PJM, where bids are provided after a 6 month delay), one may attempt a curve fit directly on the aggregate bid curve. At the time when this study was conducted, however, bid curves were not publicly available for the New England market. Instead we used the hourly price, quantity pairs published by the ISO. For each week, we preformed a least square fit of an exponential shape through the collection of 5*24 price/quantity pairs, allowing both the 'a' and 'b' parameters to vary. The estimated 'b' parameters were disregarded, while the mean of the estimated 'a' parameters was chosen as a representative supply curve shape.

6.7.2 Estimating Deterministic Seasonal Load and Supply Shapes

After the time series of historical supply states has been generated, the rest of the calibration process is virtually identical for the supply and demand components of the model. The next step is to transform the hourly time history of the load and supply states into a history of daily weights. First, however, we need to extract the deterministic seasonal components from the data. For each hour of the day, the mean load and supply level are calculated, creating the [1*24] vectors μ^L and μ^b. The process is carried out separately for each calendar month, thus creating a [12*24] mean value surface for the load and supply processes, as shown in figures 6-12 and 6-13.

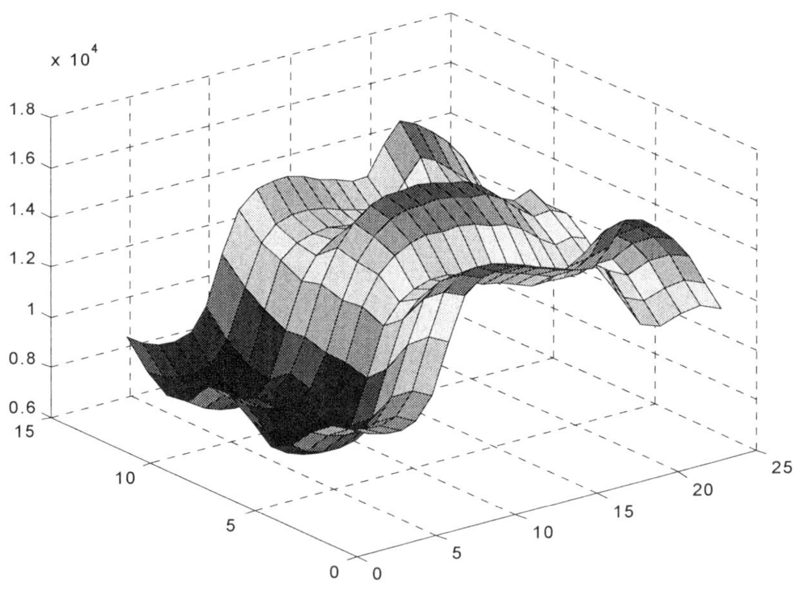

Figure 6-12 Average monthly patterns of daily load, μ_{mL} for, New England

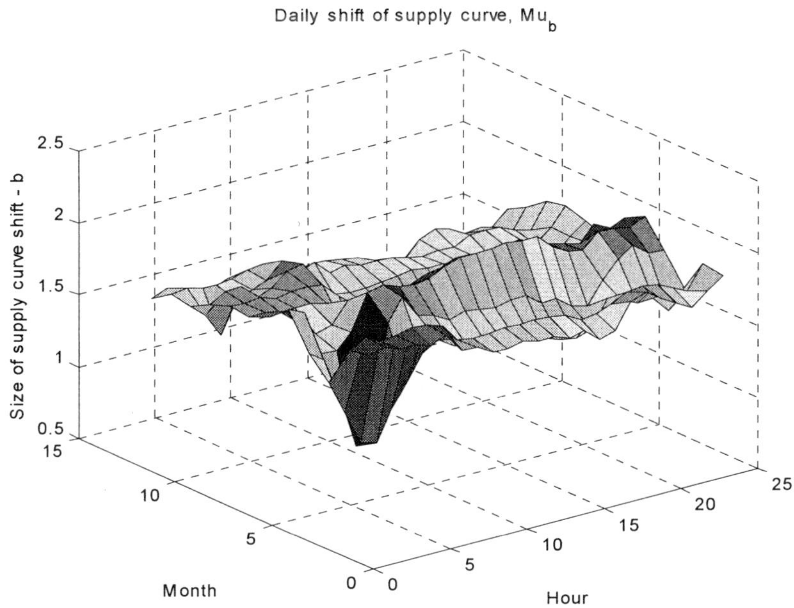

Figure 6-13. Monthly mean vector for the supply state; μ_m^b, New England

6.7.3 Calibration of Principal Component Vectors

Using the techniques of principal component analysis, as described in Appendix A, we calibrated a new set of basis vectors for the intra-day deviations in the supply and demand states from the mean surface described in the previous section. The new basis functions were calculated separately for each month, allowing for seasonality in the intra-day supply and demand dynamics.

As expected, the hourly load values inside a single month were highly correlated, so by using only the first principal component we were able to account for over 90% of the variance of the demand. The variance (in %) explained per month by the first PC is shown in table 6-2.

Month	1	2	3	4	5	6	
Var (%)	92.78	95.12	94.60	93.37	94.13	96.33	
Month	7	8	9	10	11	12	Avg.
Var (%)	96.75	96.46	93.91	94.27	93.06	93.36	94.51

Tab. 6-2: Variance of load explained by the first PC for different months

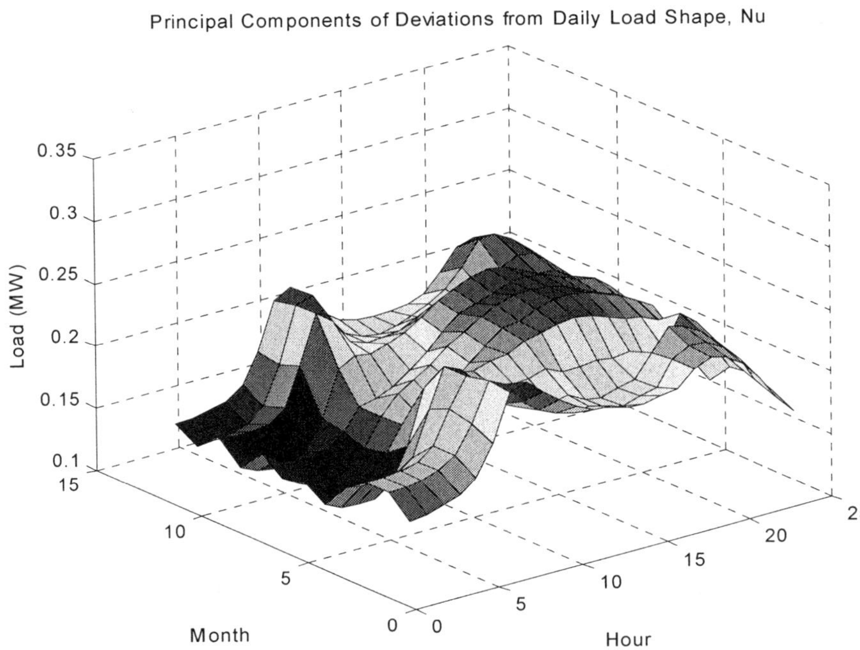

Figure 6-14. Principal Components of load $\mathbf{v}_m^L$ for New England

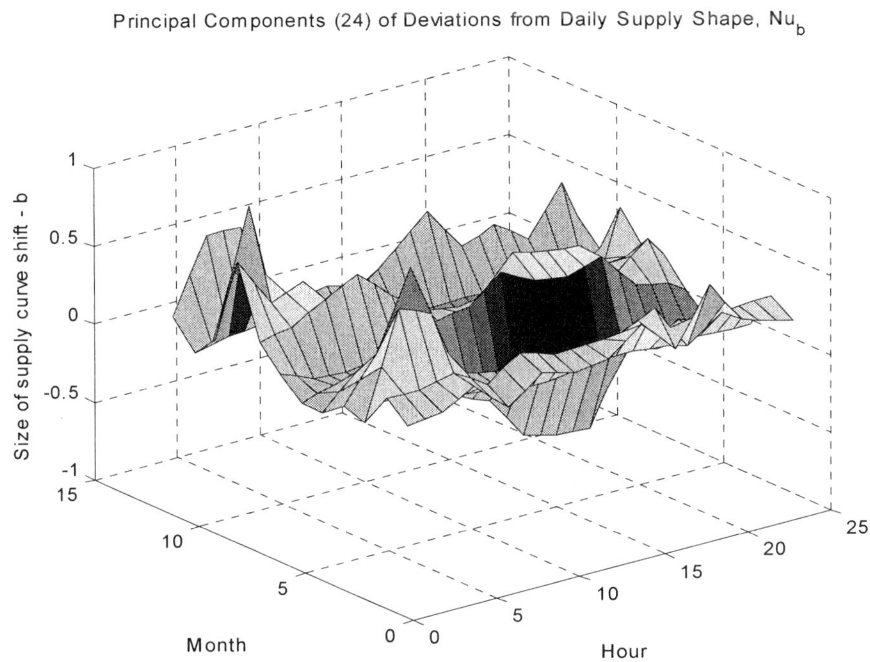

Figure 6-15 Principal Components of supply shift $\mathbf{v}_m^b$, New England

To calibrate the supply side of the BSM using the New England supply data, the problem was that we only had 14 months of hourly **b** available. To use a full-size PC Analysis, the number of instances in data (in our case workdays in a month) should be at least equal to the number of original variables, in our case the number of hours analyzed. Since on average there are only about 22 workdays in a month, we would require at least two instances for each month, raising the required number of months to 24.

To extend the available amount of data, three approaches were investigated.

1. Duplicating the missing months to obtain 24 months worth of data. Since the data was the result of two distinct stochastic processes, this would significantly alter the data beyond usability, introducing a deterministic pattern.

Treating the entire year as composed of 12 equal months, thus introducing a single set of j principal components. Since the PC analysis is used to model deviation from the monthly daily load pattern μ_m, this approach would have adverse effects on the amount of information retained by the model.

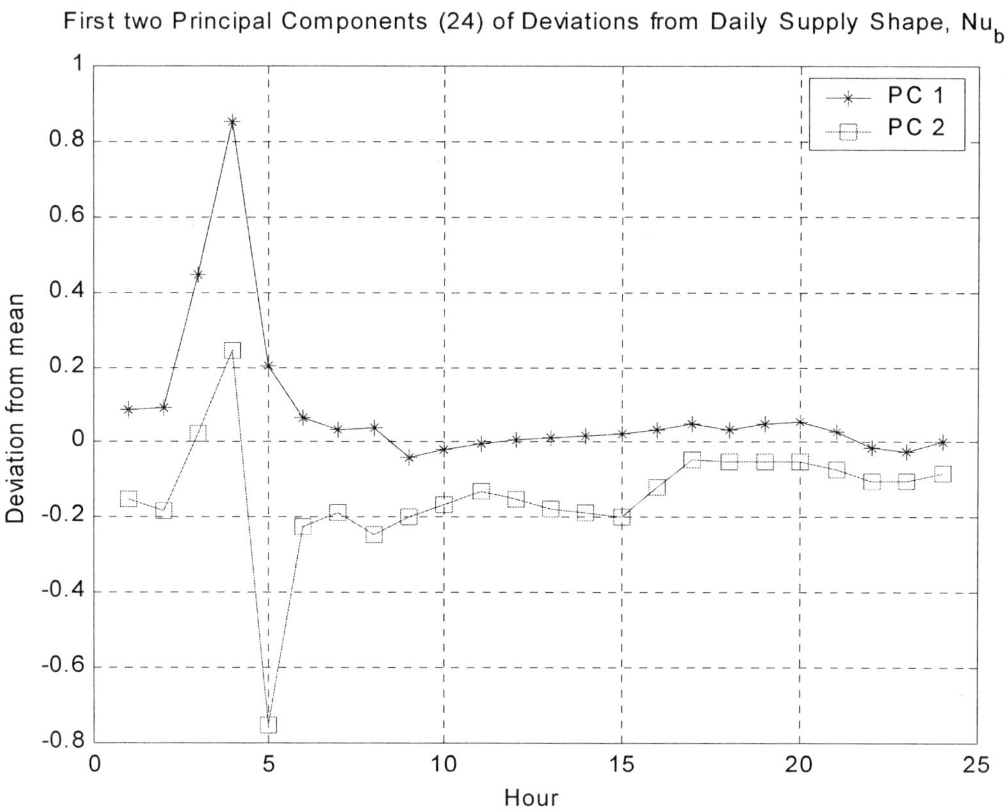

Figure 6-16 First two PCs of deviation from $\mu_m{}^b$ for 24 hours for December

2. Treating every month separately, but using a reduced number of variables to calculate the PCs. In our case, only 12 odd hours were used as original variables, reducing the order of PC matrix v to $[12 \times j \times 12]$. For the reduced order of the problem, the data for a single month (at least 20 week-days) was sufficient. After the matrix **v** was calculated and the number of retained PCs determined, the values for

the missing 12 even hours were interpolated. Table 6-3 presents the fraction of variance covered by the first five PCs.

Tab. 6-3 Variance of supply (in %) explained by the first four PCs for different months

Month PC	1	2	3	4	5	6	7	8	9	10	11	12	Avg
1	50.12	36.08	64.98	63.50	49.18	45.68	64.29	51.57	52.10	86.92	59.47	53.99	52.22
2	19.58	19.90	21.37	11.96	42.25	28.22	14.71	24.98	13.21	8.00	12.13	23.16	18.57
3	12.30	18.66	4.30	9.32	4.09	11.44	9.17	13.41	11.75	1.74	10.65	9.56	9.18
4	6.99	8.72	3.31	8.34	1.32	4.74	5.64	3.41	7.21	1.31	5.04	4.07	4.93

Here, the choice of only one PC is less obvious than in the load data. An average amount of variance explained by each of 12 original variables is 8.33 %, so according to the guidelines (Appendix A) we should have in some months considered using two or even three PCs. For the rest of this chapter, however, we will assume that a first order approximation, using a single principal component vector, gives sufficient accuracy.

6.7.4 Estimation of the Volatility, Drift and Mean Reversion Rate

The Bid-Based Stochastic Model can be expressed in the state space as:

$$\mathbf{x}_{t+1} = \mathbf{A}\mathbf{x}_t + \mathbf{B}\mathbf{u}_t + \mathbf{Q}\boldsymbol{\eta}_t$$
$$y_t = \mathbf{C}\mathbf{x}_t + \mathbf{D}\mathbf{u}_t + \mathbf{R}\xi$$

The problem of joint estimation of system and noise parameters has been solved in the literature for simpler problems; see [8],[24],[25] and [26]. However, there are significant differences between our problem and others. Some of the approaches were using a simpler two-factor model, or they assumed risk-neutrality, which does not hold true in electricity markets. In the context of financial markets, the parameters of the spot process are often calibrated using a time series of historical forward prices; see [8]. The forward curve provides a richer set of data, since it indicates the market expectation of future spot price levels for a series of maturity dates (at every time step we observe a price curve rather than a single spot price). The additional data provided by the forward markets simplifies the estimation process and eliminates the stability problem encountered when estimating from historical spot and load data (see discussion below).

The standard estimation techniques usually assume known covariance matrices of the stochastic processes in the model, i.e. process noise covariance matrix **Q** and measurement noise covariance matrix **R**. Alternatively, other techniques for estimation of noise covariances require a complete knowledge of the other system parameters. Since among the unknown BSM parameters there were also the stochastic process variances σ and σ^{δ}, the elements of the matrices **Q** and **R**, the standard estimation techniques failed to converge.

The parameters had to be estimated separately in several consecutive steps, in which the parameters were estimated independently. Since the load and the supply processes are described in a similar way in the model, their parameters α^{L}, κ^{L}, σ^{L}, $\sigma^{L\delta}$ and α^{b}, κ^{b}, σ^{b}, $\sigma^{b\delta}$ can be estimated separately and in the same way.

Estimation of the Bid-Based Stochastic Model parameters can therefore be summarized in three successive phases.

1. **Long-term drift of the mean** κ is estimated using a linear least squares fit. After κ is determined, the data is de-trended, i.e. the long-term drift is eliminated.
2. **Calculation of mean reversion factor**: Factor α, determining the mean reversion speed of the weight process, can be estimated using linear regression over de-trended data.
3. **Estimation of process volatilities**: Using the estimated α, the remaining parameters of the model in state space from σ and σ^{δ} can be estimated using the adaptive Kalman Filter and the technique for identification of the variance-covariance matrices of the process and measurement noise **Q** and **R** [30].

6.7.4.1 Calculation of the mean reversion factor

The evolution of weights can be calculated in the following way:

$$w_1 = (1-\alpha)w_0 + \alpha\,\delta_0 + \sigma\, z_0 + \kappa + \sigma^{\delta} z_0^{\delta}$$

After assuming the initial values of ($\delta_0 = 0$) and the linear trend already eliminated from the data in the previous step ($\kappa = 0$), the following sequence of equations unfolds.

$$\begin{aligned} w_2 &= (1-\alpha)w_1 + \alpha\,\delta_1 + \sigma\, z_1 + \sigma^{\delta} z_1^{\delta} \\ &= (1-\alpha)w_1 + \alpha\,\delta_0 + \sigma\, z_1 + \alpha\sigma^{\delta} z_0^{\delta} + \sigma^{\delta} z_1^{\delta} \\ &= (1-\alpha)w_1 + \sigma\, z_1 + \alpha\sigma^{\delta} z_0^{\delta} + \sigma^{\delta} z_1^{\delta} \\ w_{k+1} &= (1-\alpha)w_k + \sigma\, z_k + \sigma^{\delta}\left(\alpha\sum_{j=0}^{k-1} z_j^{\delta} + z_k^{\delta}\right) \end{aligned}$$

The last equation could be rewritten as

$$w_{k+1} = (1-\alpha)w_k + A_k$$

$$A_k = \sigma\, z_k + \sigma^{\delta}\left(\alpha\sum_{j=0}^{k-1} z_j^{\delta} + z_k^{\delta}\right)$$

Since the part of the equation denoted as A_k is influenced solely by zero-mean processes z and z^{δ}, it is a zero-mean process. It should then possible to estimate α using Linear Regression over the time series vectors of weights $\mathbf{w}_d$. The vector $\mathbf{w}_{d+1}$ is shifted in time for a day ahead compared to $\mathbf{w}_d$. The operator $\Diamond$ denotes the least-squares fit of the two vectors [27].

$$\alpha = 1 - \left(\mathbf{w}_{d+1} \Diamond \mathbf{w}_d\right)$$

There is, however, a significant problem with this approach. The variance associated with each incremental new observation grows linearly with the total number of observations, and is therefore unbounded. This violates a basic premise of the linear regression technique, which requires that the variance of the estimation error be bounded. We illustrate problem by performing a series of regression estimates on the α parameter of the load and supply process. For each consecutive estimation run we add an additional observation, using the result form the previous run as the initial guess on the parameter. A seen in figures 6-17 and 6-18, this process becomes unstable.

An interesting result was that with a limited number of samples, under 1,500 hours (about two months), the estimation technique gave reasonably stable results. The estimates for this sample region are presented in table 6-4. This property seems to indicate that there is a natural separation in the timescales at which the reverting and non reverting states evolve. In the following section we will exploit this property by proposing a time scale separated version of the Bid-based model.

Tab. 1: Estimated parameters of the BSM

	Load process	**Supply process**
α	0.3	0.75
00κ	4	-1.7e-3
a	1.13484e-4	

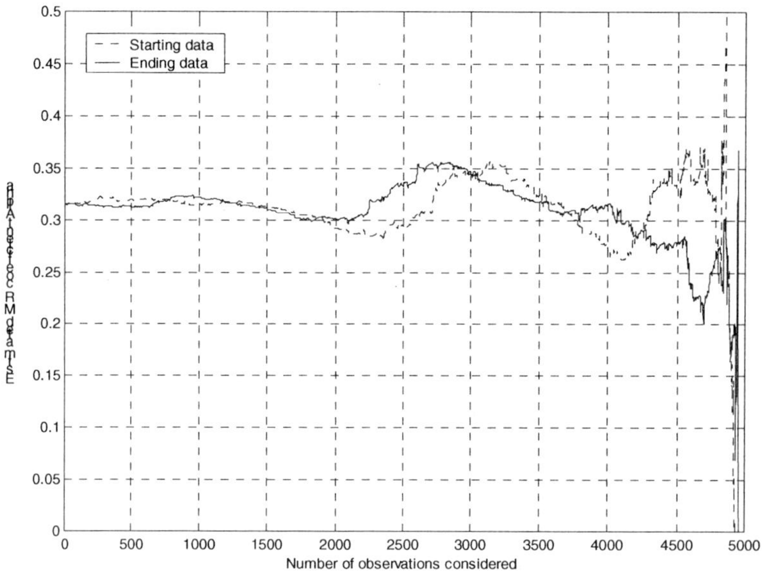

Figure 6-17 Estimation of α^L as a function of number of the samples considered

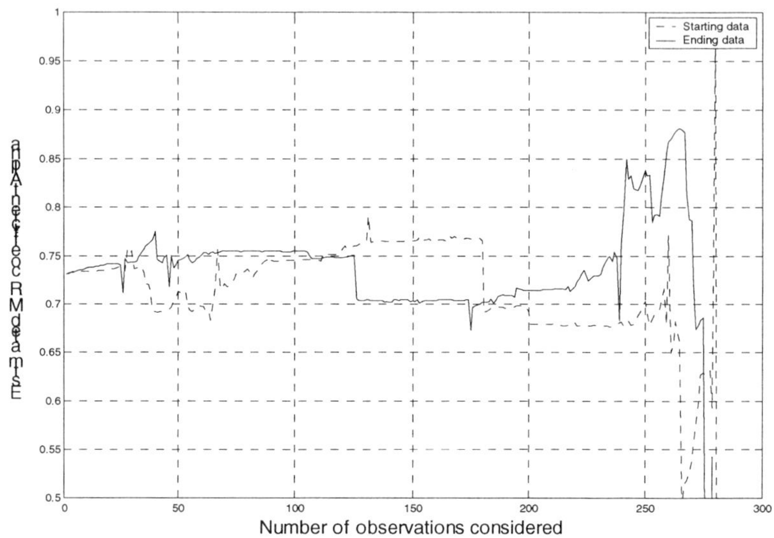

Figure 6-18 Estimation of α^b as a function of the number of samples considered

6.8 THE TIME-SCALE SEPARATED BID-BASED STOCHASTIC MODEL

In order to circumvent the problem which arises in jointly calibrating all the parameters of the load or supply process, we here introduce a variation on the model based on the notion of time-scale separation. It is based on the notion that the stochastic growth state of the supply and demand processes evolves at a slower rate than the fast dynamics. If the separation is significant enough, then the fast state will have time to revert fully back to the slow state. In the model we capture this effect by letting the long term supply and demand processes evolve on a monthly basis. The model for the dynamic evolution of the weights can then be expressed as

$$e_d = w_d - \delta_m$$

where

$$e_{d+1} - e_d = -\alpha e_d + \sigma_m z_d$$

and

$$\delta_{m+1} - \delta_m = \kappa + \sigma^{\delta} z_m^{\delta} .$$

The benefit of the time-scale separated model is that it makes the 'mean' process directly observable from the historical data. If the separation is genuine, then we can accurately approximate the monthly δ of supply or demand as the mean of the weights during this month,

$$\delta_m = \frac{1}{T_m} \sum_{\tau = T_{m-1}}^{T_m} w_{d\tau} .$$

Performing this operation on the historical data generates separate time series for the 'w' and 'δ' processes. The parameters corresponding to each factor can then be calibrated separately using linear regression techniques. The step by step process of calibrating the model is outlined below.

1. We construct a [D×1] time series vector of principal components weights w_d, **w**.

$$\mathbf{w} = [w_d], \quad d = 1..D$$

2. For every year and for every month within the current year, a mean of w_d, δ_m, was calculated. D is the total number of days in the data, while D_m is the number of days in month m.

$$\delta_m = \frac{1}{D_m} \sum_{d=1}^{Dm} w_d, \quad m = 1..12$$

$$D = \sum_{m=1}^{12} D_m$$

The [12×1] vector of monthly means δ* could then be defined as

$$\boldsymbol{\delta}^* = [\delta_m], \quad m = 1..12$$

We could also define a [D×1] vector δ, defined as a daily time series of δ_m.

$$\boldsymbol{\delta} = [\delta_{md}], \quad d = 1..D_m, \quad m = 1..12$$

3. The mean reversion of the daily weights w_d to the pertaining monthly mean δ_m is described as

$$w_{d+1} - w_d = \alpha \left(\delta_m - w_d\right) + \sigma_m z_d$$

where the change in weights is determined by the mean reversion part and the stochastic component, $\sigma_m z_d$. The stochastic process z_d is normally distributed with a zero mean and standard deviation of one.

$$z_d \approx N(0,1)$$

The mean of the stochastic process is zero and is not affected by the process volatility measure, σ_m. Coefficient α could therefore be determined using linear regression to satisfy the least-squares criterion [27].

$$y' = a + bx$$

$$b = \frac{\sum_{d=1}^{D}(x_d - \bar{x})(y_d - \bar{y})}{\sum_{d=1}^{D}(x_d - \bar{x})^2}$$

The regression is performed over the time series vectors of weights **w** and monthly means δ. A shift of the vector $\mathbf{w}_d$ for a day ahead is denoted as $\mathbf{w}_{d+1}$.

$$\mathbf{x} = [x_d] = (\mathbf{w}_{d+1} - \mathbf{w}_d)$$
$$\mathbf{y} = [y_d] = (\boldsymbol{\delta}_m - \mathbf{w}_d)$$
$$\alpha = b$$

4. Using α, a vector of the estimated weights $\mathbf{w}'_{d+1}$ was obtained:

$$\mathbf{w}'_{d+1} = \alpha\boldsymbol{\delta}_m + (1-\alpha)\mathbf{w}_d$$

5. The difference between the estimated $\mathbf{w}'_{d+1}$ and $\mathbf{w}_{d+1}$ was the contribution of the stochastic component of the process, $\sigma_m z_d$. It was therefore possible to calculate the monthly volatility measure σ_m of the process by subtracting the estimated values of $\mathbf{w}'_{d+1}$ from the actual values $\mathbf{w}_{d+1}$ and calculating standard deviation of the parts of the time series vectors, belonging to a particular month:

$$\sigma_m = StDev\left(\mathbf{w}^m_{d+1} - \mathbf{w}'^m_{d+1}\right), \quad m = 1..12$$

$$StDev(x) = \sqrt{\frac{1}{D-1}\sum_{i=1}^{D}\left(x_i - \frac{1}{D}\sum_{j=1}^{D}x_j\right)^2}$$

6. The parameters of the weight mean process δ_m have been determined using linear regression. The drift parameter κ has been calculated as a mean difference of the time shifted time series of the monthly means δ.

$$\boldsymbol{\kappa} = \boldsymbol{\delta}_{k+1} - \boldsymbol{\delta}_k$$
$$\boldsymbol{\kappa} = [\kappa_d], \quad \mathrm{d} = 1..\mathrm{D}$$
$$\kappa = ((\boldsymbol{\delta}_{k+1} - \boldsymbol{\delta}_k) \lozenge \mathbf{I})$$

7. With the help of κ, a [D×1] vector of estimated weight means δ'_{d+1} was calculated:

$$\boldsymbol{\delta}'_{d+1} = \kappa + \boldsymbol{\delta}_d$$

8. Similar to monthly volatility in **w**, the volatility measure σ^{δ} of the mean process was calculated by subtracting the estimated values of δ'_{d+1} from the actual values δ_{d+1} and computing the standard deviation:

$$\sigma^{\delta} = StDev(\boldsymbol{\delta}_{d+1} - \boldsymbol{\delta}'_{d+1})$$

After applying the algorithm to both processes, the 1st principal component's weights of load w_d^L and of supply curve shift w_d^b, a set of parameters of the model was obtained, presented in Tab., Tab. and Tab..

Tab.6-4: Calibrated parameters of the Load in TBSM

	α^L	κ^L	$\sigma_m^{\delta L}$
Load	0.0204105	75.4766	2023.60

Tab. 6-5Monthly calibrated parameters of the Supply in TBSM

	a	α^b	κ^b	$\sigma^{\delta b}$
Supply	1.13484e-4	0.0318440	-0.0536667	0.413067

Tab.6-6: Monthly calibrated parameters of the TBSM

Month	Load volatility measure σ_m^L	Supply volatility measure σ_m^b
1	876.755	0.200464
2	674.219	0.081697
3	477.651	0.350756

4	468.729	0.375337
5	831.078	0.384563
6	679.871	0.324813
7	1092.886	0.364438
8	748.013	0.104237
9	1030.650	0.177193
10	371.965	0.409969
11	779.300	0.134920
12	862.887	0.225309
Average	684.154	0.241054

A comparison of the calibrated parameters between the two versions of the model, BSM and TBSM, is shown in Tab.. The variances σ_m^L and σ_m^b in TBSM are 12-month averages.

Tab.6-7: Comparison of the estimated parameters in the TBSM and BSM

	Load process		**Supply process**	
	TBSM	**BSM**	**TBSM**	**BSM**
α	0.0204	0.3	0.0318	0.75
κ	75	4	-0.0537	-1.7e-3
σ	685		0.2411	
σ^δ	2023		0.4131	
a	1.13484e-4			

6.9 SIMULATIONS

The BSM postulates the market clearing price as an exponential function of the load and supply states.

$$S_k = e^{aL_k + b_k}$$

The dynamics of the load and supply states are described by a set of equations of the form:

$$\mathbf{X}_d = \boldsymbol{\mu}_m + \sum_{i=1}^{j} w_d^i \mathbf{v}_m^i$$

$$w_{d+1} - w_d = \alpha(\delta_d - w_d) + \sigma_m z_d$$

$$\delta_{d+1} - \delta_d = \kappa + \sigma^\delta z_d^\delta$$

Based on historical load and spot price data, parameters for both the supply and demand processes are estimated.

- **the monthly timescale**: the [24×12] matrix of average monthly 24-hour daily profile μ_m and
- **within a day on the hourly timescale**: the [24×12] matrix of monthly principal components $\mathbf{v}_m$.

These parameters are independent of the type of model we use for calculation of daily weights.

At the same time, we obtain the parameters that govern the evolution of the daily weights w_d; mean reversion speed α, long term drift κ, a [24×1] vector of daily weight volatilities σ_m and long-term mean δ_d volatility σ^δ.

Using simulation it is possible to investigate the properties of the two fundamental processes - load and supply, which drive the price in both models. The simulation also enables us to illustrate their influence on the price.

For the purpose of simulation and demonstration of the properties of the model only, the BSM volatility measures σ and σ^{δ} were approximated using the known parameters of the TBSM. Since these values don't represent true estimates, they were annotated as σ' and $\sigma^{\delta}{}'$.

- The daily weight process w_d evolves on the same timescale in both TBSM and BSM models, so their volatility measures should be roughly the same,

$$\sigma_{TBSM} = \sigma'_{BSM}.$$

- The long-term mean on the other hand develops much faster, and its volatility should therefore be much smaller, divided by a square root of the time constant factor. Assuming that there are about 25 working days in a month, the daily volatility should be about 5-times smaller,

$$\sigma'_{DVM} = \frac{\sigma_{TSVM}}{\sqrt{T}}, \quad T \cong 25 \text{ days}$$

The simulations were performed using either the Bid-based Stochastic Model or the Time-scale Separated Volatility Model, calibrated to the short-term market-clearing price. The simulations investigated the impact of different parameters for both the load and supply processes on the output of the model. The following properties were investigated:

1. Evolution of the average daily short-term market price (spot price) of electricity, as influenced by the average monthly 24-hour profiles, μ_m^L and μ_m^b, and the monthly principal components $\mathbf{v}_m^L$ and $\mathbf{v}_m^b$.
2. Expected value and Standard Deviation of the daily weights $\mathbf{w}^L$ and $\mathbf{w}^b$, driven by the daily process parameters, α, κ, σ_m and σ^{δ}.
3. Development of the daily averaged hourly price.

The results are briefly discussed and shown in the following sections.

The parameters used in simulation are given in Tab. 2. In addition to the calibrated ones, the parameters σ' and σ^{δ}' were approximated to demonstrate properties of the BSM and are presented in the shaded cells of the table.

Tab. 2: Parameters of the BSM, used in simulations

	Load process	**Supply process**
α	0.3	0.75
κ	4.0	-0.0017
σ'	685	0.24
σ^{δ}'	400	0.83
a	$1.1*10^{-4}$	

6.9.1 Monthly Parameters

The monthly average spot price in the actual data to which the model was calibrated is shown in figure 6-19 There are relatively big differences among certain months, describing a year with unusually high summer prices. The summer of 1999 was very hot and the prices were higher than the historical levels. Since only a limited amount (14 months) of price data was available, the influence of a single month in calibration of supply process was stronger than in load process calibration, where almost 20 years of data were available and the influence of excessive months were less prominent.

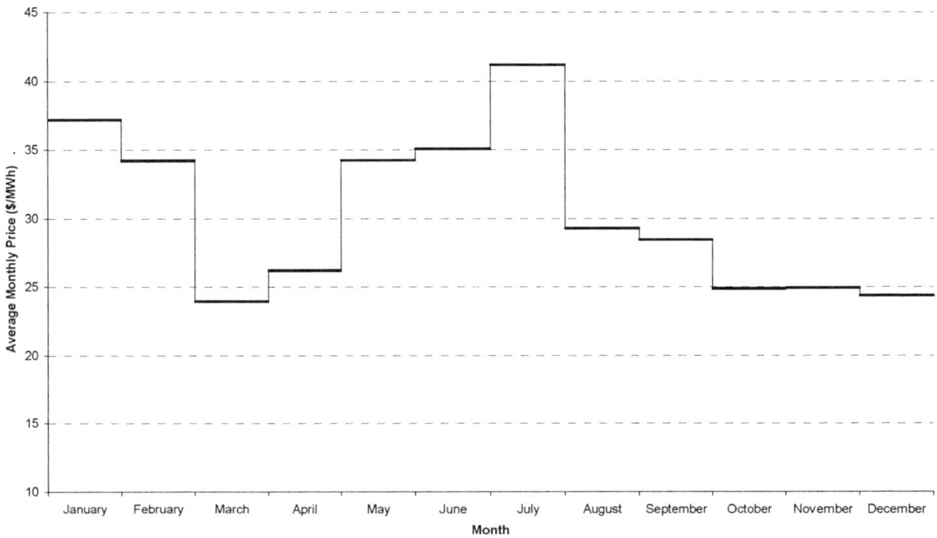

Figure 6-19 Monthly average of the spot price, New England, May '99-June '00

In Figure 6-20, the seasonal evolution of the average simulated daily spot prices are shown. The price as generated by the model exhibits similar properties as the actual average monthly price in Figure 6-19. The main difference could be observed during the summer months, where the influence of load process dampens the excessive shift in supply curve toward higher prices, as dictated by supply process.

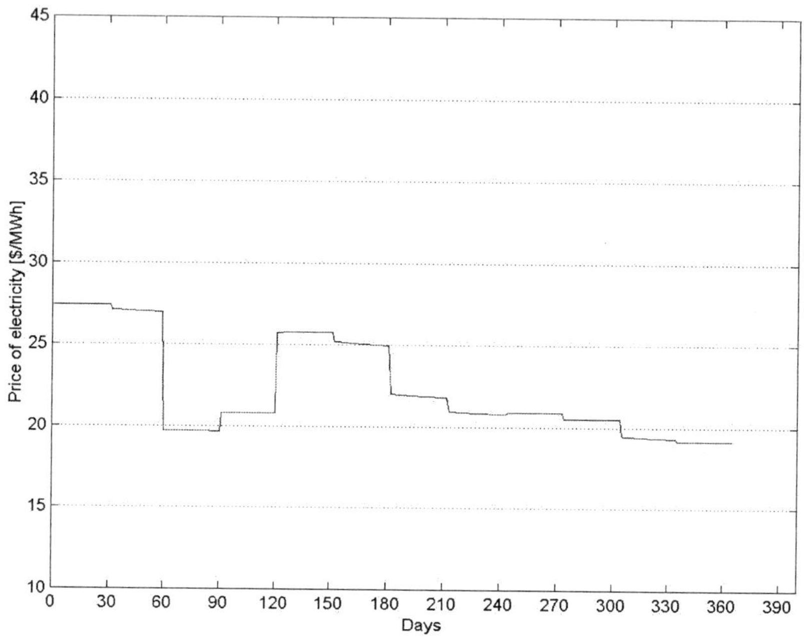

Figure 6-20: Simulated average monthly price

6.9.2 Properties of the Daily Weight Process

The stochastic properties of the model on the other hand can be illustrated without the interference of monthly mean values by examining the daily weight processes $\mathbf{w}^L$ and $\mathbf{w}^b$. They are driven by four stochastic processes - z_k^L, $z_k^{L\delta}$, z_k^b and $z_k^{b\delta}$ - and governed by the daily process parameters α, κ, σ_m and σ^δ. The interplay of the short-term $\mathbf{w}$ process variances, σ_m^L and σ_m^b, and the long-term δ process variances σ^L and σ^b in the model is shown in figure 6-21.

The short-term variances of the mean reverting process, which are bounded, dominate in the short run. As time progresses, however, the long term variances will gradually become the dominant source of uncertainty.

Similar conclusions can be drawn from Figure 6-22, where the evolution of the mean value of $\mathbf{w}^L$ and its volatility boundaries are shown. The standard deviation of the process is not uniform over the months, what is the consequence of interplay between two volatility measures, $\sigma_m{}^L$ and $\sigma^{L\delta}$. At the same time, the volatilities of electricity price differ from one month to another. During the periods of peak load, prices tend to be much more volatile than in spring or fall, which is reflected in the model output.

The mean grows steadily according to the long-term growth parameter κ^L. The weights were simulated for a two-year period with a 10.000 simulation runs.

The mean of the supply process weight, $\mathbf{w}^b$, and its volatility boundaries are shown in figure 6-22.The mean slowly drifts downwards, and the monthly shapes in standard deviation are more pronounced than in load process in Figure 6-23.

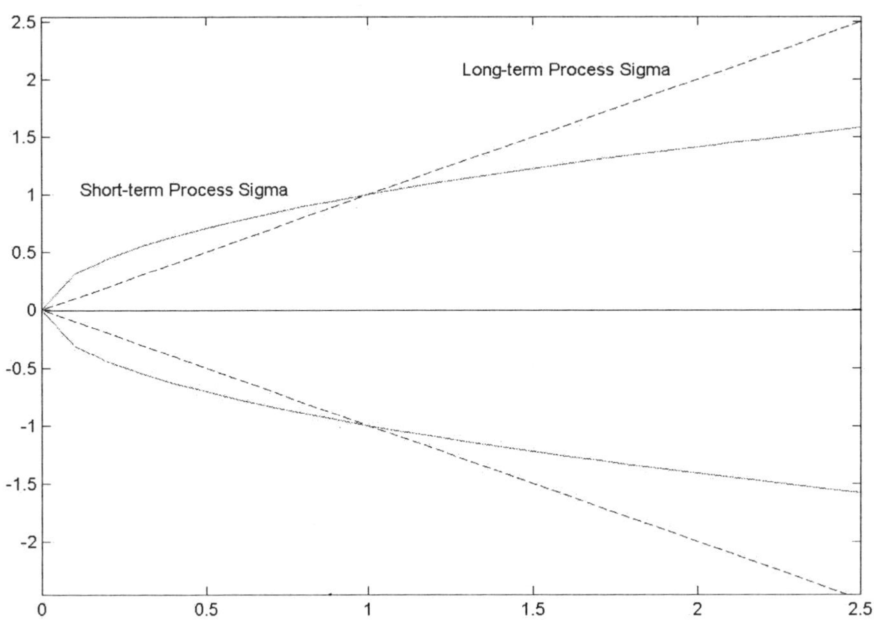

Figure 6-21 Volatilities of short- and long-term processes, σ and σ^δ

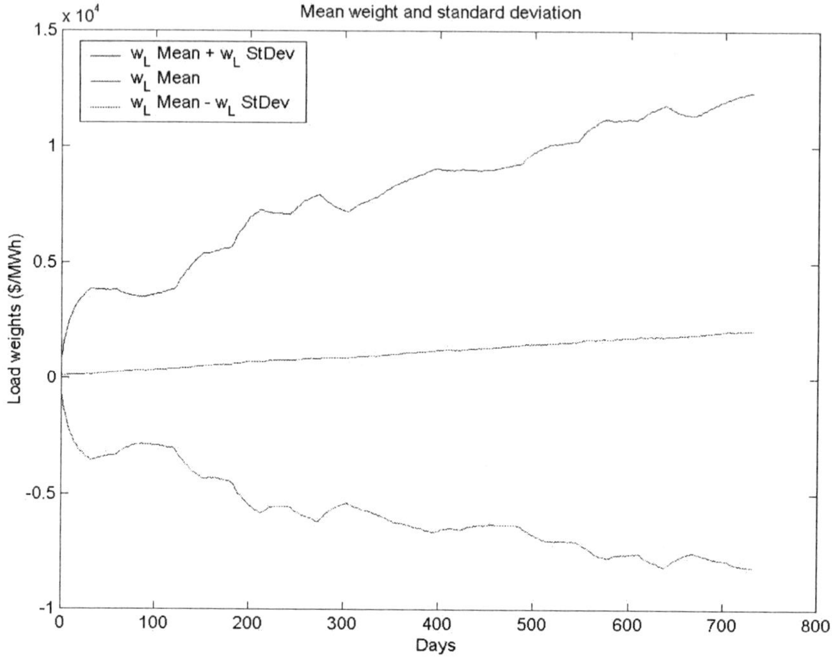

Figure 6-22: Daily weights w_m^L : mean value and standard deviation

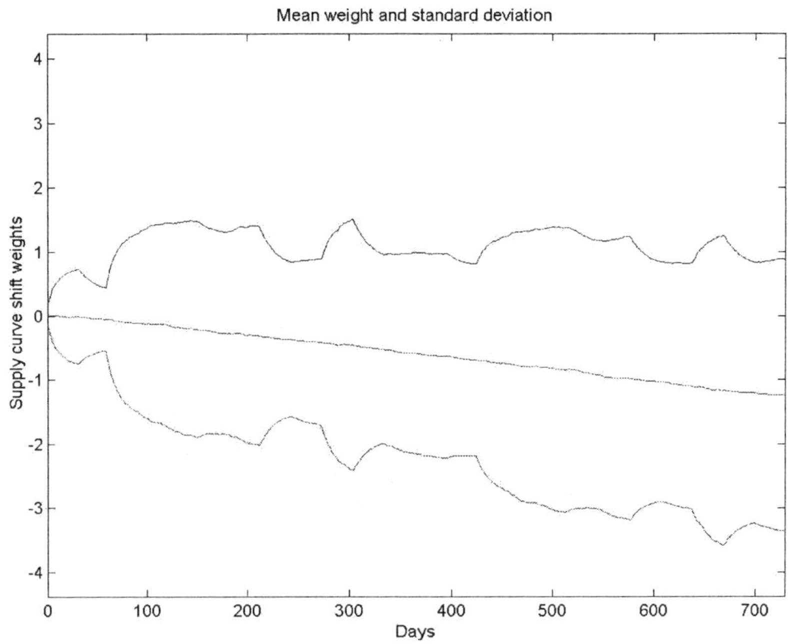

Figure 6-23: Daily weights w_m^b : mean value and standard deviation

6.9.3 Daily Price Simulations

Using the BSM it is possible to generate hourly spot price $\mathbf{S}_T$ and its volatility. Because the intra-day dynamics that can be found both in hourly development of load and hourly clearing of market in supply, it is important to have the model that is able to capture the hourly price dynamics. On the other hand, it is sometimes also necessary to neglect the hourly dynamics and deal with daily prices, as it is the case in certain applications such as forward contracts.

In the Stochastic Model the price evolves as a sequence of daily, 24-hour vectors of prices.

$$S_d = e^{a\mathbf{L}_d + \mathbf{b}_d}$$
$$S_d = [P_{dh}], \quad h = 1..24$$

The daily price would therefore be calculated as a daily average of the vector $\mathbf{S}_d$.

$$\overline{S}_d = \frac{1}{24}\sum_{h=1}^{24} S_{dh}$$

In the simulation, we have examined the development of the average price $\overline{S}_d$ during the course of one year. The mean and the standard deviation of prices were calculated in 10.000 simulation runs.

Figure 6-24 displays the mean value of electricity price and the volatility measure (in our case standard deviation) boundaries. Both prices and standard deviations exhibit strong monthly traits.

The standard deviations, when presented alone in Figure 6-25 show corresponding monthly diversity but generally agree with each other and with the observations on volatility in daily weights.

6.9.4 Daily Price Simulations with Time Scale Separation

The Time-Scale Separated Bid-based Stochastic Model properties have been investigated in the same way as the properties of the Bid-based Stochastic Model. Using the parameters from Tab. 1, the average daily price $\overline{S}_d$ was simulated for the period of one year with 10.000 runs.

The mean value of $\overline{S}_d$ is shown in Figure 6-26 together with the standard deviation as volatility boundaries. The overall shape still reacts noticeably to changes in months, while the overall impression is that the volatility is somewhat larger than in BSM. The same conclusion could be drawn from analysis of standard deviation in Figure 6-27.

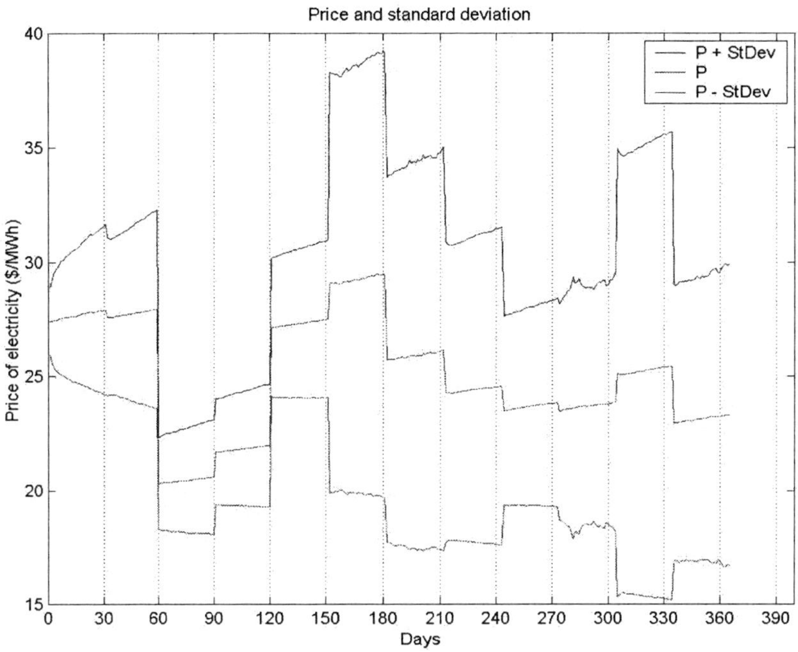

Figure 6-24 Average price $\overline{\mathrm{P}}_{\mathrm{d}}$ mean and standard deviation

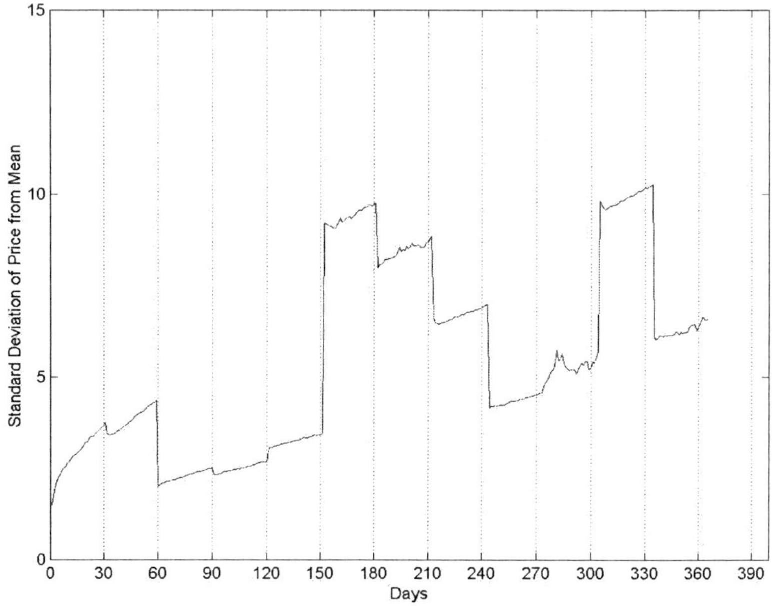

Figure 6-25: Standard deviation of the average price $\overline{P}_d$

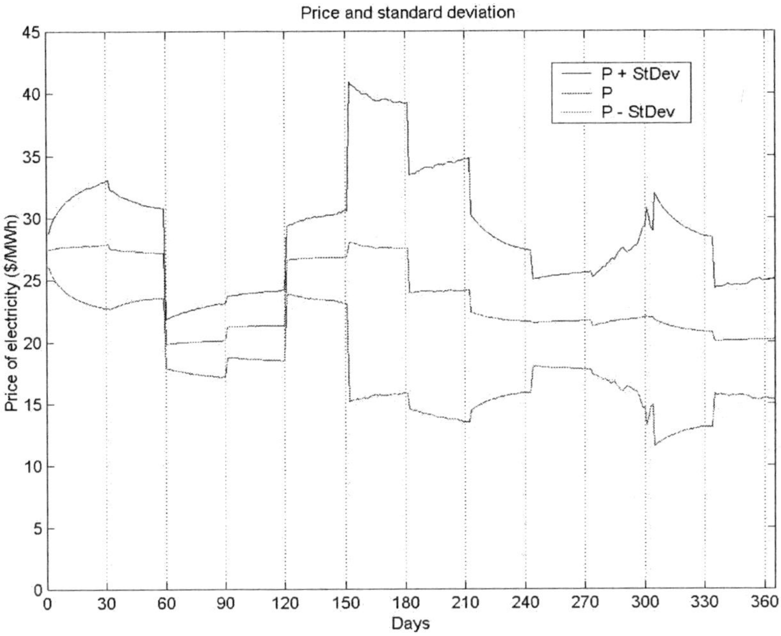

Figure 6-26: Average price $\overline{P}_d$: mean and standard deviation

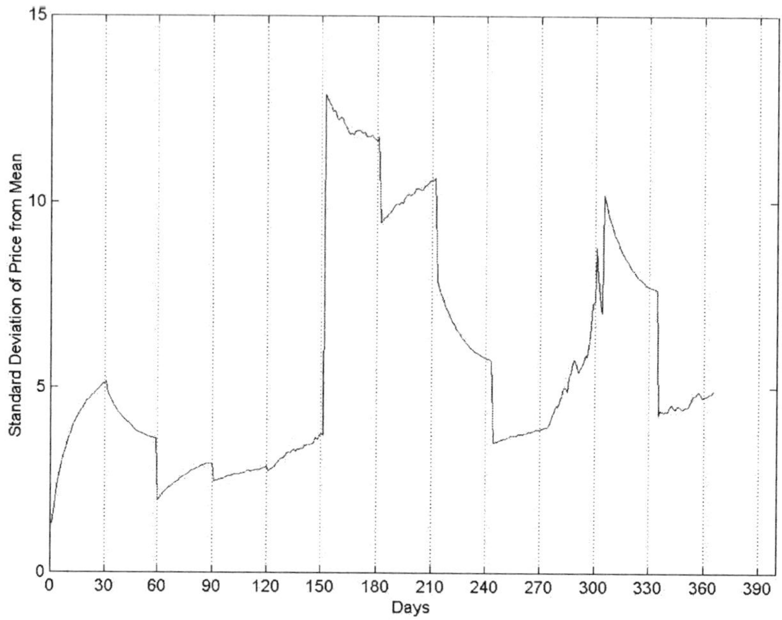

Figure 6-27: Standard deviation of the average price $\overline{P}_d$

6.10 CONCLUDING REMARKS

A wide range of literature has evolved on the modeling of electricity markets, and on the associated price dynamics. Here we have briefly reviewed a number of approaches: quantitative, cost based, economic equilibrium, agent based, and experimental, all with their own advantages and drawbacks. The modeling approach used is generally dependent on the type of problem addressed. A marketer may use one type of model for optimizing the short term bidding of assets, another for hedging in the forward markets, yet a third model for evaluating investment decisions. Ultimately, the user would like all of these models to reflect the best current estimate of the future. However, given the range of modeling approaches, it is often hard to check whether the models used are consistent with each other.

The bid-based model presented in this report is intended as a fundamental model for electricity price dynamics, and is to be used in a wide range of applications. The emphasis was placed on incorporating the unique characteristics of electricity prices, including seasonality on multiple time scales, lack of load elasticity, stochastic supply outages, strong mean reversion, and stochastic growth of load and supply.

The second emphasis was on reducing the computational complexity of the model. This was achieved by applying techniques such as principal component analysis, which reduced the dimensionality of the model drastically, with a minimal loss in performance.

A timescale-separated version of the model was calibrated on real market data (New England). The lack of price data for the market makes the calibration on the supply side of the model tentative, but as more data becomes available, the parameter estimates will become firm. The scheme for calibrating the original version of the model is outlined, but its implementation is left to future research.

Chapter 7

Optimal Futures Market Strategies for Energy Service Providers

The previous chapter focused on developing a stochastic model for the dynamics of electricity prices. Here we show how the bid-based model can be applied to help market participants optimize their decision making under uncertainty. This chapter will address the problems facing energy service providers. In Chapter 8 we address the supply side of the market.

7.1 HEDGING RISK FOR ENERGY SERVICE PROVIDERS

With the deregulation of the electric utilities came a significant increase in the financial risk to the companies, known as load serving entities (LSE) or energy service providers (ESP), serving the end users. The risk faced by the energy service provider can be traced back to the physical and economic interactions between the ESP and its customers. The physical configuration of the distribution network does not generally allow for any differentiation in the service provided to different customers, in terms of power quality or reliability. Economically, the ESP is limited by its contracts to serve retail customers, typically known as standard offer contracts, which are structured to shield the customers from any fluctuation from the wholesale price of electricity. The financial risk from the wholesale market must therefore be absorbed by the ESP. In this chapter we break down the sources of this risk, and propose a methodology for the ESP to hedge its customer portfolio using a dynamic futures trading strategy.

The key to effective risk management for load serving entities is to understand the temporal dynamics of the uncertainty in the market price (price risk), and in the customer demand (quantity risk), as well as the correlation between the two. The cash flow for a load serving entity is defined over a given delivery period as a function of uncertain future price and demand levels. Next we impose dynamic constraints on the price and load levels using the bid-based model developed in the previous chapter. The risk preference of the firm is characterized using a mean-variance formulation. The trading strategy consists of a decision rule determining the quantity of forward contracts to purchase or sell at given times prior to the delivery period, as a function of observed price and demand levels. In this book two cases are examined. In the static hedging formulation, the ESP is allowed to trade a single time prior to the delivery period. The decision problem in this case becomes a nonlinear programming formulation, and is solved through simulation. Next we extend the problem to the dynamic hedging case, where the firm is allowed to trade at fixed intervals prior to the delivery period. To pose this problem one must extend the price model to incorporate uncertainty in the dynamics of the forward price. The problem suffers from a case of the curse of dimensionality. The computational complexity grows quickly as a function of the number of delivery periods, the number of hedging intervals, and the number of state variables in the underlying spot, forward and load models. We examine properties of the models which could allow for simplifying assumptions, as well as a possible reformulation of the objective function which could reduce the computational intensity of the problem.

7.2 THE PHYSICAL AND ECONOMIC INTERACTION OF ENERGY SERVICE PROVIDERS AND THEIR CUSTOMERS

Many of the problems inherent in the physical and economic interaction of ESPs with their customers can be traced back to a paradigm developed by society with regard to the industry during decades of strict regulation. Electricity is considered by many as a common utility, even a social right, rather than a commodity. Implicit in this viewpoint is that the providers of electricity should not be free to withhold their product, or lower the quality of their service, merely for economic reasons. The old regulatory structure did little to discourage such an attitude. Since returns for the utilities were essentially cost based, and there was no such thing as a bad customer or a bad investment. The objective of the utilities during this period is best described by the phrase 'keeping the lights on.' Whatever the cost of ensuring that the lights did indeed stay on, the firm could be certain to recover its expenditures.

In the deregulated environment, the idea that the customer has a 'right' to electricity does not correspond to the objectives of the energy service providers. Energy service providers are profit driven entities, with an obligation to their share holders to extract the maximum profit possible from the provision of electricity to retail customers. In some instances, this may entail not serving a customer if the cost of providing the service exceeds the willingness of the customer to pay. This is where the historical role of utilities catches up with the deregulation process. Since customers were traditionally considered to have a right to service, and furthermore all customers had the right to the same service (at least at the retail level), the distribution infrastructure was not built to accommodate customer segregation of any type. The two major deficiencies in the current distribution system are the metering and interruptability of customers.

1. **Metering:** Electricity meters generally record only the aggregate consumption of electricity by the customers. Furthermore, in most areas the meters are not connected to the ESP thorough modems or other communication devices, but must be read manually. As a result, the only information available to the ESP is the monthly aggregate consumption by each customer. As described in the previous chapter, electricity exhibits significant daily and weekly price cycles. As a result, the cost of serving two customers who use the same number of KWh's a week, but have different patterns of consumption, may differ significantly. In order to measure the real cost of serving a customer, the ESP needs to measure the users power profile; that is, the typical rate of consumption for each hour of the day. This information can be used in order to provide

a fair price of service, in which customers do not cross-subsidize each other.

2. **Interruptability:** The non-storability of electric power requires that a delicate balance between the rate of production and consumption be upheld at all times. At certain times, either due to unusually high demand levels, or due to the unexpected loss of generation resources, there may not be sufficient generation available to meet the instantaneous load. In such cases, the system must shed load in order to avoid a system breakdown. Currently there are no means for the energy service provider to curtail selected customers who may be willing to offer such a service in return for reduced rates [33]. Instead they have to rely on area wide rolling blackouts (as experienced recently in California). Such ad hoc measures are clearly damaging to the overall social welfare since they do not distinguish between how different customers value the reliability of the service.

The lack of installed technology to handle metering and selective interruptability currently prevent the ESPs from providing differentiated services to various groups of customers. This has effectively prevented the establishment of a competitive retail market for electricity [34]. It leaves the energy service providers holding inflexible standard offer contracts with significant price and quantity risks. In fact one could argue that, since they offer identical services, the only area in which ESPs can compete with each other is in the ability to manage this risk.

7.3 PROBLEM FORMULATION

We consider the situation where an energy service provider has obligated itself to serve a group of customers at a fixed rate. The length of the contract is generally several years, and the service provider has no means of 'opting out' of the contract. Furthermore, the contract is structured in such a manner that the customers may consume as much or as little power as they want at any time without additional penalties. This setup is similar to the current standard offer contracts being offered to retail electricity customers. Furthermore we impose the constraint that the ESP owns no generating assets but purchases all power from the spot market. This exposure to the spot market leaves the ESP with significant financial risk. To mitigate this risk it can purchase financial futures contracts on electricity through the commodities exchange. We will here address the problem of how to generate an optimal trading strategy for the LSE in the futures market.

7.4 MODELING

We will now generate mathematical models necessary to pose the hedging problem for the energy service provider [36]. The general components of this problem were outlined in Chapter 2, and they include modeling the price (and quantity) uncertainty, incorporating this uncertainty into an overall cash flow model, and optimizing with respect to the firms risk preference.

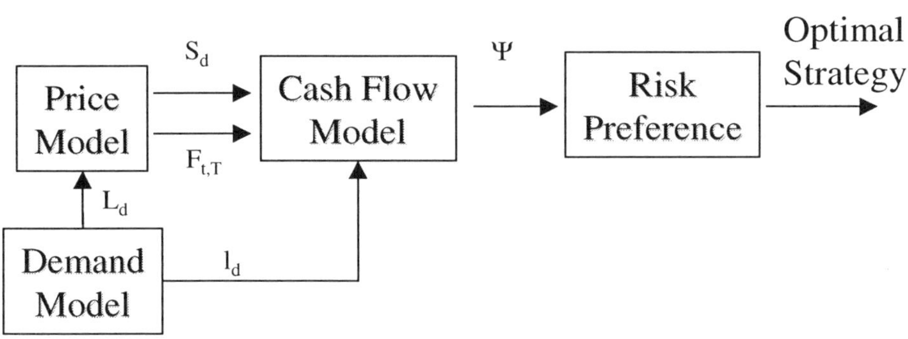

Figure 7-1

The variables and parameters of the model are defined in Table 7-1.

R	Fixed rate for customer (standard offer) ($/MW)
S_d	Spot price for day d.
l_d	Total amount consumed in day d (MW)
$F_{ti,T}$	Price of a forward contract for delivery in month starting at T, as seen at the time of purchase t_i ($/MWh)
$q_{ti,T}$	Total quantity of forward contracts purchased for delivery month starting at T (MW/h) at time of purchase t_i.
M	total number of delivery days in a month
N	number of months in the delivery period
T	Starting day of delivery period
t_0	Starting day of hedging period
K	Number of rebalancing intervals in the hedging period

Table 7-1 Variables and parameters in the ESP problem formulation

The time period during which the ESP has committed to serving its customers is broken down into N delivery periods, each spanning one month. This is done in order to accommodate the structure of the futures market. Recall that each traded futures contract requires delivery for one month. We will furthermore make the simplification of using a single load and price variable per day. In reality there are 24 hourly spot prices in a given day. At the end of the chapter we discuss how the modeling approach can be extended, using principal component theory, to account for intra-day variations

7.4.1 Cash Flow Model

We begin by modeling the cash flow for the LSE before any purchases in the forward market, which we call the unhedged cash flow (Ψ^U). The total Ψ^U is given by

$$\Psi^U = \sum_{m=1}^{N} \sum_{d=Mm+1}^{M(m+1)} l_d (R - S_d)$$

For simplicity we will here consider the case where the hedging period is a single month. The cash flow function then becomes:

$$\Psi^U = \sum_{d=T}^{T+M} l_d (R - S_d)$$

Next we consider the cash flow incurred from a portfolio of forward contracts $q_{tj,T}$. This is the cash flow of the hedge (Ψ^H).

$$\Psi^H = \sum_{d=T}^{T+M} \left[\sum_{j=0}^{H} q_{t,T} (S_d - F_{j,T}) \right]$$

The index t represents the time at which the forward contracts are purchased or sold. In contrast to the spot market, which clears at discrete daily intervals, the forward market trades in real time. For computational purposes, however, we will restrict trading to discrete intervals t_j. Each t_j represents a hedging interval. Changing the number of futures contracts held from one interval to the next is known as rebalancing the portfolio, or rolling over the hedge. Since a forward contract $F_{t,T}$ cannot be purchased after the starting date of the delivery month (T), we include the constraint $t_i \leq T$. The timeline of the hedging process is shown in Figure 7-2.

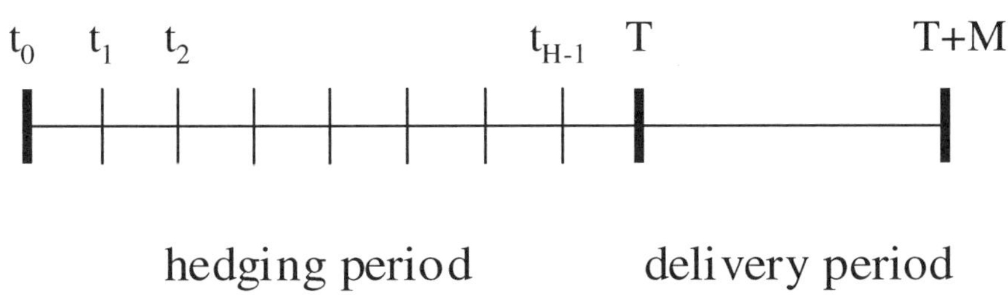

Figure 7-2: Time line for dynamic hedging problem

The hedging period $[t_0,T]$ is divided into H hedging intervals of equal length. The number of hedging intervals used will generally depend on the transaction cost and liquidity in the forward market. The total cash flow for the hedged LSE (Ψ) can be expressed by

$$\Psi = \sum_{d=T}^{T+M}\left[l_d (R - S_d) + \sum_{j=0}^{H} q_{t_j,T} (S_d - F_{t_j,T}) \right]$$

Note that the forward contracts have no cash flow prior to the delivery period (they are not pre-paid), so that all cash flow occurs during the delivery period. Ψ is therefore equal to the net profit of the ESP with respect to the load serving contract.

7.4.2 Price and Demand Models

The load level, as well as the spot and forward prices at any future date, are random variables in the cash flow formulation. To characterize the distribution of these variables we apply the bid-based stochastic mode developed in the previous chapter. Since future values of the forward price occur as explicit terms in the cash flow, we must extend the price model to incorporate the dynamics of the forward price.

7.4.2.1 Bid-Based Model

The model for the spot price is identical to the one developed in Chapter 6, except that we are applying a daily rather than an hourly model, and are ignoring the unit outage component.

Spot Price Model:

$$S_d = e^{aL_d + b_d}$$

Load Model:

$$\mathbf{L}_{\mathrm{d}} = \boldsymbol{\mu}_{\mathrm{m}}^{\mathrm{L}} + \mathrm{w}_{\mathrm{d}}^{\mathrm{L}},$$
$$e_{d+1}^{L} - e_{d}^{L} = -\alpha^{L} e_{d}^{L} + \sigma_{m}^{L} z_{d}^{L}$$
$$\delta_{d+1}^{L} - \delta_{d}^{L} = \kappa^{L} + \sigma^{L\delta} z_{d}^{L\delta},$$

where

$$e_{d}^{L} = w_{d}^{L} - \delta_{d}^{L}.$$

Supply Model:

$$\mathbf{b}_{\mathrm{d}} = \boldsymbol{\mu}_{\mathrm{m}}^{\mathrm{b}} + \mathrm{w}_{\mathrm{d}}^{\mathrm{b}}.$$
$$e_{d+1}^{b} - e_{d}^{b} = -\alpha^{b} e_{d}^{b} + \sigma_{m}^{b} z_{d}^{b}$$
$$\delta_{d+1}^{b} - \delta_{d}^{b} = \kappa^{b} + \sigma^{b\delta} z_{d}^{b\delta},$$

where

$$e_{d}^{b} = w_{d}^{b} - \delta_{d}^{b}.$$

7.4.2.2 Forward Price Model

We model the forward price as the expectation of the average spot price over the delivery period of the contracts. In addition we allow for the presence of a seasonal risk premium λ. The risk premium can be positive or negative depending on the market, and furthermore is allowed to be seasonal in the maturity date of the contract.

$$F(t,T)=e^{\lambda(T-t)}E_t\left\{\frac{1}{N}\sum_{T}^{T+N}S_d\right\}$$

We model the risk premium of the forward risk premium so that there it grows linearly with the log of the average spot price. This makes sense since the log of the spot price corresponds to the sum of the load and supply states. Recall from our discussion of the properties of the bid-based model that the long term component of the load and supply states (corresponding to the δ processes) has a variance which grows linearly in time. In essence the model proposes that the market assign a premium to the forward contract in proportion to the variance of the underlying asset. While this makes intuitive sense, there is not yet sufficient data available to test this assumption properly. A more general model may assign a stochastic process to the risk premium.

7.4.2.3 Demand Model

One of the challenges of the load serving problem is to understand the correlation between the price and the demand for electricity. Demand is not generally price elastic, as consumers do not adjust their behavior based on the market price (or even observe prices for that matter). However since short term demand deviations are mainly temperature driven, the same inputs which drive the total market load is likely to also effect the customers served by the ESP. The bid-based model explicitly captures the effect of market demand fluctuations on the spot price. What remains to be modeled is the relationship between the system load and the ESP customer demand. In this chapter we will assume the ESP serves a rough cross section of the demand in the market, and will model the ESP demand (l_d) as a fraction of the total load (L_d), plus a noise term

$$l_d = \beta L_d + \sigma^l z_d^l .$$

7.4.2.4 Model Properties

The model can be characterized in state space format, where the state vector is given by

$$x_d^T = \lfloor w_d^L \quad \delta_d^L \quad w_d^b \quad \delta_d^b \rfloor,$$

the control variables as

$$u_d = [q_{d,T}]$$

a vector of disturbances

$$z_d^T = \lfloor z_d^{L,d} \quad z_m^{\delta L,b} \quad z_d^l \rfloor,$$

a vector of output variables

$$y_d^T = \lfloor S_d \quad l_d \quad F_{d,T} \rfloor$$

and a vector of time varying parameters

$$\theta_m^T = \lfloor \mu_m^{L,b} \quad \kappa^{L,b} \quad \alpha^{L,b} \quad \sigma^{L,b,l} \quad \sigma^{\delta L,b} \quad \lambda \rfloor.$$

The state dynamics are linear, while the output variables are a nonlinear function of the states. We can write the model in compact form

$$x_{d+1} = Ax_d + Bu_d + Qz_d$$
$$y_d = f(x_d, z_d, \theta_m).$$

7.4.3 Modeling the Firm's Risk Preference

So far we have defined the cash flow function and the dynamic constraints on the underlying stochastic variables. Now we must define the firm's risk preference, in order to arrive at an objective function for the optimization problem. The mean-variance formulation defines a firm's utility in terms of a linear tradeoff between the expected value and variance of the payoff:

$$U_d(\psi) = E_d\{\psi\} - r * \operatorname{var}_t\{\psi\}.$$

7.4.4 Summary of the Hedging Problem

The firm needs to develop a decision rule, or policy, for the quantity of forward contracts to purchase at each hedging interval (q_{tj}), in order to maximize its objective function:

$$\max_{q_{tj}} J_{t_0}$$

$$J_{t_0} = E_{t_o}\{\Psi\} - r * \operatorname{var}_{t_0}\{\Psi\}.$$

The cash flow Ψ is defined as

$$\Psi = \sum_{d=T}^{T+M}\left[l_d(R - S_d) + \sum_{j=0}^{H} q_{t_j,T}(S_d - F_{t_j,T}) \right],$$

and is subject to the dynamic constraints,

$$x_{d+1} = Ax_d + Bu_d + Qz_d$$
$$y_d = f(x_d, z_d, 6_m),$$

for the set of state and output variables defined above.

7.5 EFFICIENT REFORMULATION OF COST FUNCTION

The optimization problem defined above seems suitable for a dynamic programming (DP) solution algorithm [40],[41]. There is a set of well-defined dynamic state equations, coupled with a cost function based on expected future payoffs. When attempting to fit the problem into the standard dynamic programming form, however, we run into several problems. First, the policy (or control) of the firm is defined over the hedging interval $[t_o, T]$. The payoff or cost function, on the other hand, is defined over a non-overlapping delivery interval [T, T+N]. The DP algorithm, however, requires that the incremental cost at each stage, $g_k(x_k, u_k)$, be defined, allowing us to write the objective function as a sum of the cost incurred at each stage,

$$J_q(x_0) = E\left\{ g_N(x_N) + \sum_{t_o}^{T} g_k[x_k, u_k(x_k), w_k] \right\}.$$

To transform the problem into one with an additive cost function, we create a value function V_t, which measures the expected cash flow during the delivery period, as seen from each step in the hedging period,

$$V_{t_i} = E_{t_i}\{\Psi\},$$

or in expanded form,

$$V_{t_i} = E_{t_i}\left\{ \sum_{d=T}^{T+M} \left[l_d(R - S_d) + \sum_{j=0}^{T} q_{t_j,T}(S_d - F_{t_j,T}) \right] \right\}.$$

Next, the mean variance formulation is superimposed upon the new value function. V_{t0} represents the initial expected return of the unhedged portfolio. The objective of the hedging strategy is to maximize the expected value of the portfolio while minimizing the risk (or variance) of the return. We define the mean variance objective function as,

$$\max_{q_{ti,T}} J = E\{V_T - V_{t_0}\} - rVar\{V_T - V_{t_0}\}.$$

The new objective function differs from the old in that it does not optimize over the variance of the actual cash flow. Instead we attempt to maximize the risk adjusted value of the expected cash flow up to the end of the hedging period.

Based on the stochastic models described in the previous section, we can describe the properties of the model and the value function V_d. The model is Markov, meaning that all information about future outputs is contained in the current values of the state. As a result the value function V_d is also Markov. Furthermore V_d is a Martingale process,

$$E_d(V_{d+\tau}) = V_d \, .$$

The proof for this is a simple application of iterated expectations.
Using these characteristics of the value function, we can now rewrite the objective function of the hedging problem as

$$J = E\{V_T - V_0\} - rVar\{V_T - V_0\} \, .$$

7.6 SOLUTION APPROACHES

With the reformulated cost function, the problem would seem to fit into a dynamic programming framework. This however is not the case. The penalty on the variance of the value makes the problem impossible to solve through backward iteration. Specifically, it can be shown that the Bellman Equation is violated [38], thus invalidating the dynamic programming approach. There have been results, however, illustrating that variance-penalized Markov decision problems can be solved through nonlinear programming. Methods presented in [37],[38],[39] and [42] present such techniques for a variety of interpretations of variability. The discussion is usually in terms of infinite horizon problems, but can be generalized to the finite horizon case.

7.7 THE END STATE PROBLEM

Under the objective function defined above, the firm optimally manages the change in the value function V_t up to the final stage in the hedging period (t=T). In reality, however, the firm is concerned with the uncertainty of the cash flow occurring over the delivery interval [T, T+N]. Assume that the firm has executed an optimal policy over the hedging interval. At time T, the firm then holds a portfolio of forward contracts defined by $q=[q_{t0}\ q_{t1}\ \dots\ q_{T-1}]$. Conditional on the q vector, the ESP can then solve a single stage decision problem with respect to the original optimization problem to find the optimum number of forward contracts to buy or sell in the last stage:

$$\max_{q_T} J_T = E_T\{\Psi\} - r * \mathrm{var}_T\{\Psi\}$$

$$\Psi = \sum_{d=T}^{T+M}\left[l_d(R - S_d) + \sum_{j=0}^{H} q_{t_j,T}(S_d - F_{t_j,T})\right].$$

The end state problem has several features which make it easier to solve than the original problem.

- The problem is a single stage decision problem.
- All forward prices, past and current, are known, so $F_{t,T}$ are deterministic variables.
- Since we are at the beginning of the delivery month, the long term states of the price and load models ($\delta^{L,b}$), which evolve at a monthly rate, can be assumed to be known constants. This reduces

the underlying dynamic constraints from a fourth to second order model.

The end state problem becomes a nonlinear programming problem with dynamic constraints. Using the reduced order of the model as suggested above, the problem can be solved through brute force simulation. The two components of the objective function, the mean and variance of the cash flow, can be characterized separately as a function of the decision variable q_T. The expected return is simply a linear function of q, with a slope determined by the risk premium,

$$E\{\Psi\} = q_T\left(F_{T,T} - E_t\left\{ \frac{1}{N}\sum_{T}^{T+N} S_d \right\} \right) + K ,$$

where K is a constant. The variance of the cash flow as a function of the decision variable is quadratic skewed by the summation, as shown for the sample plot below.

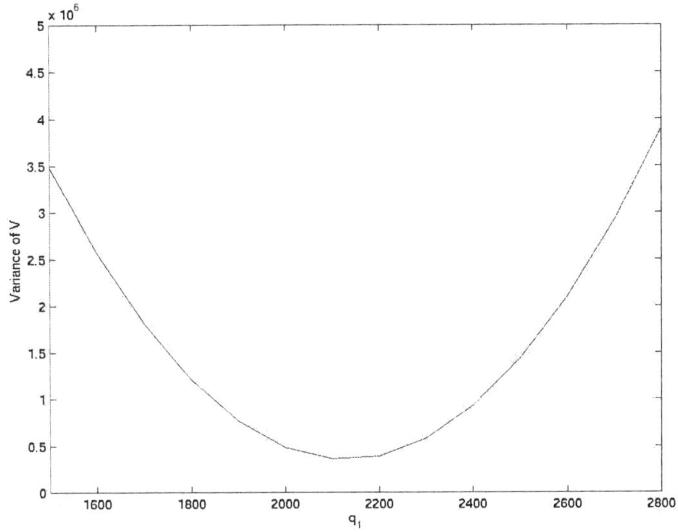

Figure 7-3: Cash flow variance as a function of the number of future contracts held.

7.7.1 Static Optimization Over Multiple Delivery Periods

In the above discussion we have limited the problem to a single delivery month. In reality the ESP is likely to have customers signed up from multiple years, corresponding to dozens of delivery periods. This adds significant computational complexity to the optimization problem. While we will not here attempt to solve the general multi-delivery period problem, we illustrate some of its properties by considering the end state problem with two delivery periods. This can be thought of as a simple hedging strategy for an ESP who does not trust the dynamic forward price models. The optimization takes place over a single stage, so that the relevant forward prices are observable. Furthermore, since the formulation is static in the decision process, the constraints on the objective function can be loosened. In this case we use the value at risk formulation in order to illustrate a different approach to risk management. Value at risk emphasizes the probability of the firm suffering a critical loss over the delivery period. Figure 7-4 shows the probability of such a loss as a function of the quantity of forward contracts bought or sold in each of the two delivery periods (q_1 and q_2).

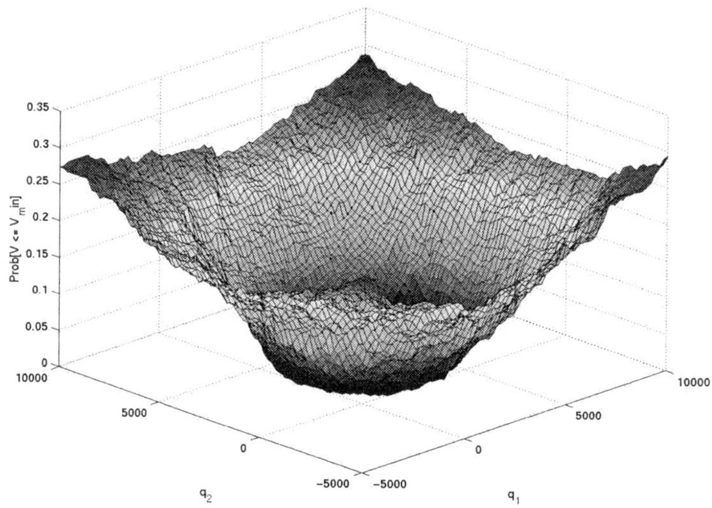

Figure 7-4: Probability of a Critical Loss as a Function of Future Portfolio

Next a constraint is imposed on the maximum allowable probability of the critical loss. This divides the space of possible futures portfolios into an admissible and an inadmissible region, as shown in Figure 7-5.

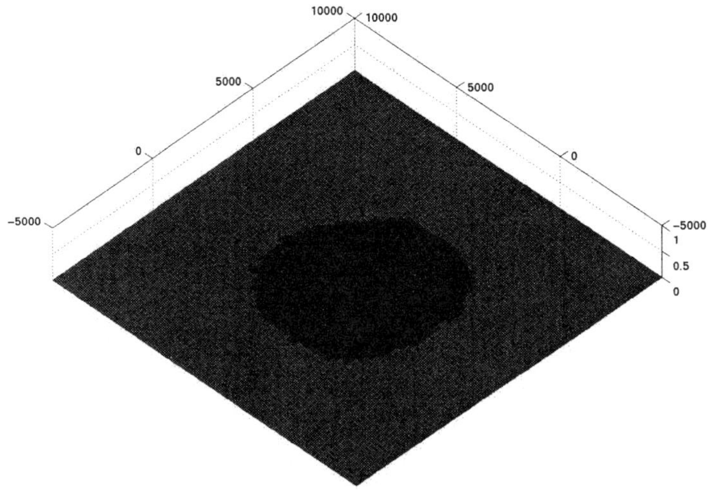

Figure 7-5: Admissible region for futures trading under VAR

The above figures were generated by brute force simulation of a range of possible future portfolios. Such an approach would be extremely time-consuming for greater numbers of delivery periods. However the convex form of the convergence region lends some hope that more efficient search techniques can be used in order to arrive at an optimal trading portfolio under the value at risk assumption.

7.8 THOUGHTS ON THE COMPLEXITY OF THE ESP HEDGING PROBLEM

Under realistic circumstances, the energy service provider hedging problem quickly escalates in dimensionality and, simultaneously, in the computational complexity of the solution. The dimensionality of the problem is governed by:

1. The number of states/sources of uncertainty in the underlying dynamic constraints.

2. The number of times the futures portfolio is rebalanced.
3. The number of months in the delivery period.

Each of these factors represents a challenge to overcome in order to arrive at an efficient formulation of the problem. Further research is needed to arrive at ready-for-use algorithms. We here provide some thoughts on possible approaches for reducing the impact of the sources of complexity.

As implied in the discussion of the end state sub-problem, the multiple time scales contained in the bid-based model, can be exploited in order to reduce the order of the model. The load and supply states each have a fast component (w) evolving on a daily time scale, and a slow component (δ) evolving on a monthly scale. When dealing with uncertainty in the near term (within the next month), the long term component can be assumed to be constant. When addressing an uncertainty projected several months into the future, the short term uncertainty can be assumed to have reverted back to the long term mean, and thus ignored. In each case the fourth order model is reduced to a second order model. The cutoff between short and long term is dependent on the mean reversion rate of the demand and supply processes, and can be best understood by examining the implied term structure of the model. For further discussion on the efficient implementation of the hedging algorithms, see [43].

In determining the proper structure of the hedging period, the user has to consider two questions. How far in advance does one need to start hedging the cash flow at delivery, and how often must the portfolio be rebalanced? The first question is limited by the forward market, which currently only trades fifteen months into the future. The second part is constrained by transaction costs, as well as by the computational complexity of the DP problem. Given a fixed number of times the user can rebalance the portfolio over a long hedging period, it may useful to adopt a nonlinear function for the length of each rebalancing period. For a mean reverting process, the further away in time the user is projecting the uncertainty, the less new information will enter over a given time interval. Consequently, the further away from the delivery period, the less likely it is that the portfolio has deviated from its optimal value. Further research is needed to find an optimal function for the length of the hedging interval as a function of time to delivery.

Chapter 8

Valuing Generation Assets

8.1 INTRODUCTION

As the electric utility industry becomes more competitive, the question of how to value generation assets becomes critical. This problem is typically approached by defining the generator in terms of its efficiency (heat rate) in converting fuel to electricity. Based on this rating, the valuation is performed by modeling the generator as a spread option between the price of the fuel used and the price of electricity [44]. The payoff from such an option is given by

$$CF_k = \max\{P_k^e - C(P_k^f), 0\}$$

where P_k^e is the price of electricity and C is the cost marginal cost of production as a function of the fuel price P_k^f.

The problem with this formulation is that it ignores several important constraints involved in the operation of the unit, such as start-up and shut-down costs, minimum run time, and maximum ramp rate [45]. These constraints have a significant effect on how units are bid into, and dispatched by, a spot market operator, and therefore have a significant effect on the owners' cash flow. By ignoring the unit commitment constraints, one is likely to undervalue plants with significant flexibility (such as micro turbines and fuel cells) while overvaluing large inflexible fossil plants.

The reason why the unit commitment problem is often ignored in the valuation of a power plant can be linked to computational complexity. In general the operator of the unit has to solve a complex dynamic programming problem to arrive at the optimal unit commitment decision for the generator [7]. This is a computationally intensive problem with polynomial growth over the time horizon over which the optimization is carried out. Therefore while it is feasible to solve the unit commitment decision for a day-ahead bidding problem [45], it is extremely challenging to extend this notion to a multi-year valuation problem.

In this chapter, we propose a new method for valuing generation assets with unit commitment constraints, using a principal component based model for spot price described in detail in [19]. The effectiveness of the principal component representation comes from being able to determine the hourly prices within each day. This is qualitatively different from using a single daily spot price, which does not recognize intra-day price variations. By applying principal components, we are able to define today's net profit from the generation asset as a function only of today's and yesterday's average spot price. By storing the mapping from the state of the spot price to the cash flow of the generator in a lookup table, we are able to simulate generator profits over multiyear periods with minimal computational complexity [46].

Next we introduce a stochastic model for the fuel price. This adds a third dimension to our lookup table, but still allows us to simulate the cash flow for the generator with a computational time growing linearly with the length of the valuation period.

The principal component based model is applied next to the problem of hedging generation assets. By defining the daily cash flows from the unit as a derivate of fuel and electricity prices, we can derive optimal strategies for trading in fuel and electricity forward markets in order to manage generation risk.

8.2 A PRINCIPAL COMPONENT BASED PRICE MODEL FOR ELECTRICITY SPOT MARKETS

The price model used is a simplified version of the bid-based price model introduced in [19]. We define a daily [24*1] price vector P_d^e, whose elements are the 24 hourly electricity prices. Next we define the log of the price vector to be the sum of a deterministic and stochastic component. The deterministic component is composed of a monthly vector μ_m, which captures the seasonal characteristics of the electricity spot price. The stochastic component is modeled as the product of the principal component vector v_m and a daily stochastic scalar weight w_d. The principal component captures the shape of daily price variations from the seasonal mean μ_m while the weight describes the magnitude of the deviation as well as its correlation over time. The log of the vector of spot market prices can be written as

$$\ln(P_d^e) = \mu_m^e + w_d^e v_m^e$$

Next we model the process describing the evolution of the weights w_d^e. We choose a two-factor discrete time mean reverting process. This captures important features of electricity markets such as short-term mean reversion and long term stochastic growth. For an in-depth discussion of modeling electricity prices, please see [19].

$$e_{d+1}^e - e_d^e = -\alpha^e (e_d^e) + \sigma_m^e z_d^e$$
$$\delta_{d+1}^e - \delta_d^e = \kappa^e + \sigma_m^{\delta e} z_d^{\delta e}$$

where,

$$e_d^e = w_d^e - \delta_d^e .$$

The form of the price process postulates that hourly spot prices will be log-normally distributed. Furthermore prices inside of a day are perfectly correlated, since they are a function of a single daily random variable w_d. This reduction in complexity is made possible by choosing the principal component in an intelligent manner.

8.2.1 Formulating the Unit Commitment Decision

To calculate cash flow in a one-day period, firstly a generator solves a unit commitment problem in order to determine when a unit is turned on or off optimally. The one-day cash flow is the expected sum of profits from operating in each hour. Let $CF_d(x_0)$ be the cash flow on day d:

$$CF_d(x_0) = \max_{\pi} E\{\sum_{k=0}^{24-1} P^e_{d,k} q - c_k(q)\}$$

Rewrite this optimization in a Dynamic Programming (DP) framework, adopted from [7] as the following:

$$CF_d(x_0) = J_0(x_0) = \max_{\pi} E\{\sum_{k=0}^{23} u_k (P^e_{d,k} q_k - c_k(q_k, P^f_{d.k}) -$$
$$I(x_k < 0)S) + (1 - u_k) \cdot (c_f + I(x_k > 0)T\}$$

$$J_N(x_N) = 0$$

Reward-to-go in hour k:

$$J_k(x_k) = \max_{u_k} E\{u_k (P^e_{d,k} q_k - c_k(q_k, P^e_{d,k}) -$$
$$I(x_k < 0)S) + (1 - u_k) \cdot (c_f + I(x_k > 0)T) + J_{k+1}(x_{k+1})\}$$

$$x_{k+1} = \begin{cases} \max(1, x_k + 1) & u_k = 1 \\ \min(-1, x_k - 1) & u_k = 0 \end{cases}$$

$$u_k \geq I(1 \leq x_k \leq t_{up})$$

$$u_k \geq 1 - I(-t_{dn} \leq x_k \leq -1)$$

$$q_{min} \leq q_k \leq q_{max}$$

$$I(\text{'TRUE'}) = 1$$

$$I(\text{'FALSE'}) = 0$$

The optimal policy is applied to obtain the maximum expected profits. The above problem is a full-blown version in which there are multiple sources of uncertainties. The first one is from electricity spot prices, and the second one is from fuel prices.

8.2.2 Price Model Used in the Unit Commitment Problem

When solving the unit commitment bidding problem, we use an approximate version of the price process. In the full-blown model we can write the next days weight as a function of today's states:

$$w_{d+1}^{e} = (1-\alpha^{e})\, w_{d}^{e} + \alpha^{e}\delta_{d}^{e} + \kappa^{e} + \sigma^{e} z_{d}^{e} + \sigma^{\delta e} z_{d}^{\delta e} .$$

We here assume that $\alpha^{e} << 1$, and $\sigma^{e} >> \sigma^{\delta e}$. We only use this assumption when formulating the day-ahead bidding strategies. With this assumption we can write the weight process as:

$$w_{d+1}^{e} = w_{d}^{e} + \kappa^{e} + \sigma^{e} z_{d}^{e} .$$

This effectively states that in the very short term (day-ahead) we can ignore the mean reversion as well as the long term volatility. It should be noted that we only use this assumption to arrive at a bidding strategy. When simulating future spot price for valuation purposes we use the full-blown version of the price model.

Therefore, we can simplify the above unit commitment by assuming that

1) $q_k = q_{max}$,
2) in a one-day period, $P_{d.k}^{f}$ is given $\forall k$,

3) $\alpha^e = 0$, during a one-day period,

$$\tilde{w}_{d+1} = \tilde{w}_d + \sigma_m^e z_d^e$$

Therefore,

$$J_k(x_k, w_{d-1}, \alpha^e) = \max_{u_k} E_{\tilde{w}_d | \tilde{w}_{d-1}} \{u_k(P_{d,k}^e q_k - c_k(q_k) \\ - I(x_k < 0)S) + (1 - u_k) \cdot (c_f + I(x_k > 0)T) + J_{k+1;}(x_{k+1}, w_{d-1}, \alpha^e)\}$$

and,

$$CF_d(x_0, \tilde{w}_{d-1}, \alpha^e) = J_0(x_{0,} \tilde{w}_{d-1}, \alpha^e)$$

8.3 CREATING A LOOKUP TABLE OF CASH FLOWS

We have shown how to calculate cash flow and a on/off policy for a generator in a given day d. This cash flow is an expected cash flow, given $\tilde{w}_d = \tilde{w}_{d,i}$. $\tilde{w}_{d,i}$ is a sample value of $\tilde{w}_d$, which is continuously normally distributed with ($\tilde{w}_{d-1} + \kappa^e$) mean and standard deviation σ_m^e. To create a lookup table mapping a pair $(\tilde{w}_{d,j}, \tilde{w}_{d-1,i})$ to a cash flow, we generate $\tilde{w}_{d,j}$ given $\tilde{w}_{d-1,i}$; and apply the optimal policy $\pi(\tilde{w}_{d-1,i})$ to determine the cash flow in the d period with $\tilde{w}_{d,j}$, or $P_{d,j}^e$. Therefore, in period d, we have

$$CF_d(\tilde{w}_{d-1.i}, \tilde{w}_{d,j}) = \sum_{k=0}^{23} \{u_k(\tilde{w}_{d-1,i})(P_{d,k}^e(\tilde{w}_{d,j}) \cdot q - c(q) - I(x_k < 0)S) \\ + (1 - u_k(\tilde{w}_{d-1,i})) \cdot (c_f + I(x_k < 0)T)\}, \forall j$$

Repeating this process using $\tilde{w}_{d-1,i}$ for other i, one obtains a cash flow matrix which an element (i,j) is a cash flow associated with both $\tilde{w}_{d,j}$ and

$\tilde{w}_{d-1,i}$. This matrix captures possible cash flows in a given month m with a simplified price process for each day d.

$$\overline{CF}_m = \begin{bmatrix} CF_d(\tilde{w}_1, \tilde{w}_1) & CF_d(\tilde{w}_1, \tilde{w}_2) & \cdots & CF_d(\tilde{w}_1, \tilde{w}_N) \\ & \ddots & & \\ \vdots & & CF_d(\tilde{w}_i, \tilde{w}_j) & \vdots \\ & & \ddots & \\ CF_d(\tilde{w}_N, \tilde{w}_1) & & \cdots & CF_d(\tilde{w}_N, \tilde{w}_N) \end{bmatrix}$$

For each month m, a cash flow matrix can be calculated by using the same method. Note that each cash flow matrix is obtained by assuming a constant fuel price $P^f_{d,k}$ for all k and d within each month m.

$$\{\overline{CF}_1, \overline{CF}_2, \ldots, \overline{CF}_{12} \mid P^f_{d,k} = \theta^f_1\}$$

8.3.1 Incorporating Stochastic Fuel Prices

Next we propose a model which will allow us to include stochastic fuel prices. We here assume that we are dealing with a gas-fired plant, but the model is equally applicable to oil or coal. Since gas is a storable commodity, it experiences less short-term volatility, and very little intra-day volatility. We will therefore make the following assumptions.

1. There is only a single daily gas price $P^f_{d,k}$.
2. In the unit commitment decision, the day-ahead gas price is assumed to be forecasted with near-perfect accuracy (i.e. assumed to be deterministic).

Next we postulate a model for the daily gas price. The log of the price is written as the sum of a deterministic seasonal and a stochastic component.

$$\ln(P_d^f) = \mu_m^g + w_d^e$$

Note that we do not need to apply the principal component approach since the price is a scalar. The stochastic component is described by a two factor mean reverting model.

$$e_{d+1}^f - e_d^f = -\alpha^f (e_d^f) + \sigma_m^f z_d^f$$
$$\delta_{d+1}^f - \delta_d^f = \kappa^f + \sigma_m^{\delta f} z_d^{\delta f}$$

where,

$$e_d^f = w_d^f - \delta_d^f$$

To apply this model we first need to expand the lookup table to include gas price as a third dimension. Note that the assumption that unit commitment takes the day-ahead gas price as deterministic allows us to add only one rather than two dimensions to the lookup table.

We now generate simulated paths for future w's for electricity and gas. The lookup table converts them into paths of future cash flows.

To capture the dynamic of fuel prices, which is assumed to vary on a monthly basis, we create a set of cash flow matrices with different fuel prices, obtaining

$$\begin{cases} \{\overline{CF}_1, \overline{CF}_2, \ldots, \overline{CF}_{12} \mid P_{d,k}^f = \theta_1^f\} \\ \vdots \\ \{\overline{CF}_1, \overline{CF}_2, \ldots, \overline{CF}_{12} \mid P_{d,k}^f = \theta_L^f\} \end{cases}$$

Note that we can generate N samples of w of fuel prices to create an $(N \times N \times N)$ cash flow matrix for each month m. This matrix completely captures uncertainty due to both the electricity and fuel prices.

8.4 LINKING SIMULATED PRICES TO THE LOOKUP TABLE TO GENERATE SIMULATED CASH FLOWS

Once the lookup table has been created, we can use the full-blown price model to generate simulated weights. The lookup table is then used to generate a path of cash flows from a path of weights. The simulation time is linear in the length of the valuation period. Furthermore we are not restricted to the proposed model for generating weights. The lookup table can be linked to any stochastic model which produces weights for the principal components.

8.4.1 Generation Asset Valuation

The value of a generator V is the expected sum of discounted cash flows during the period of valuation,

$$V = E\{\sum_{d=0}^{D} (r)^{d} \cdot CF_{d}(w_{d}^{e})\},$$

where r is a discounted factor, in which $0 < r < 1$, and D is a period of valuation (such as a 15-year period or a 15*365-day period).

There are two valuation methods that we consider here:

1) A Monte Carlo Simulation Method

  a) *M* paths of (w^{e}, w^{d}) are generated.

$$\begin{bmatrix} \overline{w}^{e} \\ \overline{w}^{g} \end{bmatrix}^{i} = \begin{bmatrix} w_{1}^{e}, w_{2}^{e}, \ldots, w_{N}^{e} \\ w_{1}^{f}, w_{2}^{f}, \ldots, w_{N}^{f} \end{bmatrix}^{i};$$

b) For each path i of $[w^e, w^f]^i$, each cash flow is obtained by choosing a cash flow associated with each pair of $[w_j^e, w_j^f]^i$ from the lookup table. The value of a generator if w_j^e and w_j^e follow path *i* is the sum of the discounted cash flows.

$$V^i = \sum_{d=0}^{D} (r)^d \cdot CF_d^i(w_d^e, w_d^f)$$

c) The value of a generator is equal to

$$V = \frac{1}{M}\{\sum_{i=1}^{M} V^i\} = \frac{1}{M}\{\sum_{i=1}^{D}(\sum (r)^d \cdot CF_d^i(w_d^e, w_d^f))\}$$

2) A Multinomial Tree Method

Instead of using Monte Carlo simulations to model uncertainties in the spot price, we can use a multinomial tree method. The simplest version of the tree method is a binomial tree. A binomial tree can be used when only one source of uncertainty exists. For example, if gas prices are known at any time d, the only uncertainty in generation asset valuation comes from electricity prices. At the end of each day d, $w_d^{e,i}$ either goes up with probability p to be $w_{d+1}^{e,i+1}$ or goes down with probability 1-p to be $w_{d+1}^{e,i-1}$. Hence, at each node a cash flow associated with w_d^e ($w_d^{e,i+1}$ or $w_d^{e,i-1}$) conditioned on the previous node $w_{d-1}^{e,i}$ is simply obtained from the lookup table. We expand the tree from a single node on day 0 to 2^N nodes on day N. The value at each node i on day d V_d^i is the expected sum of discounted cash flows of the next adjacent nodes, plus the cash flow associated with $w_d^{e,i}$ incurred at that node.

$$V_d^i = CF_d^i(w_{d-1}^{e,j}, w_d^{e,i}) + r(p \cdot CF_{d+1}^{i+1} + (1-p) \cdot CF_{d+1}^{i-1}), j = \{i\text{-}1, i\text{+}1\}$$

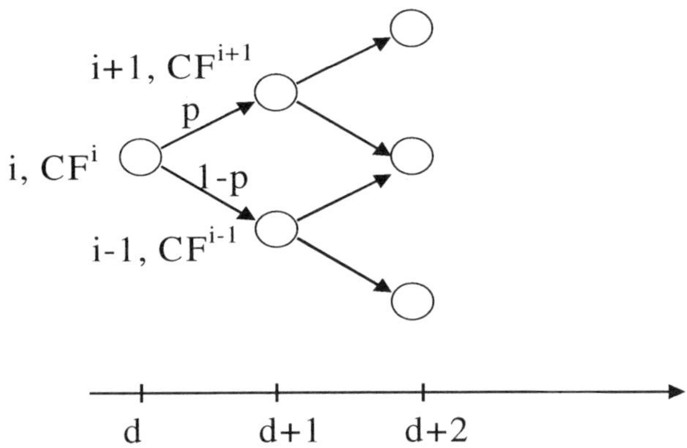

Figure 8-1. A Binomial-Tree Representation

For a case with more than one sources of uncertainty, two additional branches will be added to capture each additional source of uncertainty. This will make the problem become more complex since nodes grow exponentially with time. A Monte Carlo approach might be more applicable to deal with more than one source of uncertainty.

8.5 CONCLUDING REMARKS

As shown in this chapter, the day-ahead process of deciding on an optimal commitment strategy for generation assets with unit commitment constraints under uncertain fuel prices is an extremely complex problem. While in theory one can value a unit by simply extending this commitment decision for a multiyear period, by simulating a range of possible fuel and electricity price paths, this approach is extremely computationally demanding. The use of principal component theory in price modeling recognizes that there are dominant patterns in the hourly price deviations within a day. These patterns can be exploited to reduce the number of random variables in the unit commitment decision, and greatly reduce the computational complexity of the problem. We have illustrated how we can characterize the day-ahead bidding strategy of a generator as a function of three states, representing electricity and fuel prices. This leads to the creation of a lookup table which effectively stores all information related to the unit commitment problem. Once this lookup table is created, the problem of valuing the generator is trivial, since all we have to do is generate simulated price paths for fuel and electricity prices. The approach is extremely computationally efficient and allows us to simulate the value of generation assets over a multi-year period.

The contribution here is in presenting a computationally feasible valuation method, which will allow users to differentiate between technologies based on flexibility as well as fuel efficiency. The authors believe that this will have significant impact on investment choices in the new industry, both on a wholesale and distribution level.

8.6 FIGURES

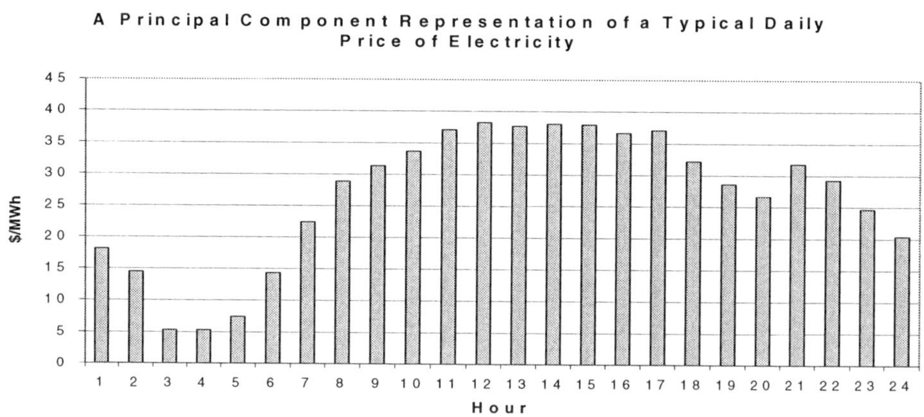

Figure 8-2. Principal Component Representation of a Typical Daily Price of Electricity[1]

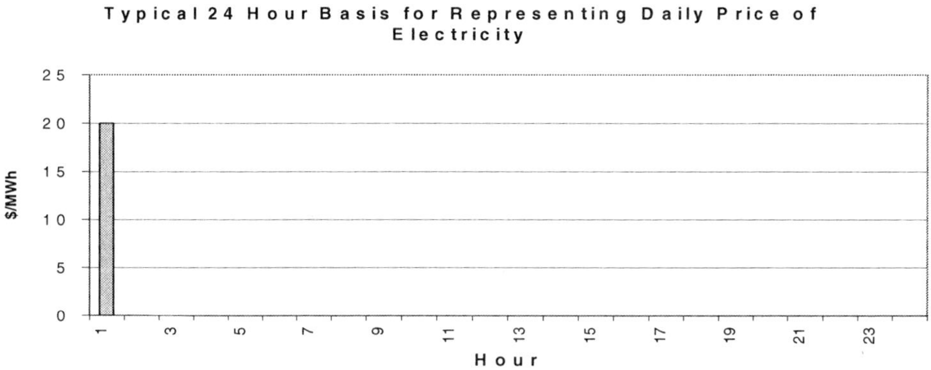

Figure 8-3. A Typical 24 Hour Basis for Representing the Daily Price Vector for Electricity (Hour 1)

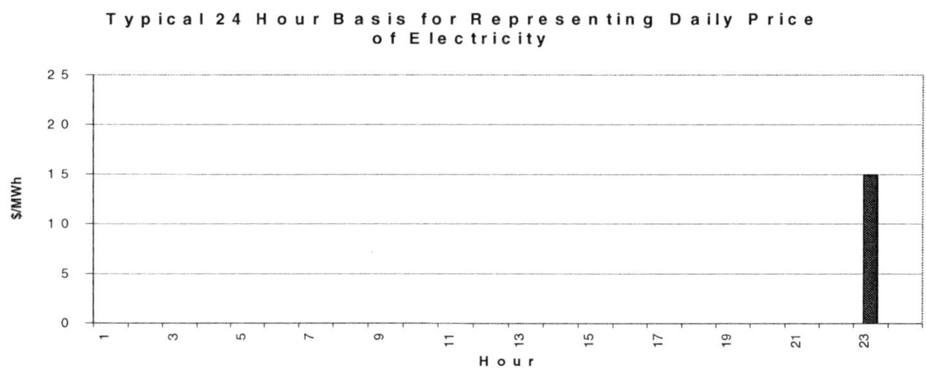

Figure 8-4 A Typical 24 Hour Basis for Representing the Daily Price Vector for Electricity (Hour 23).

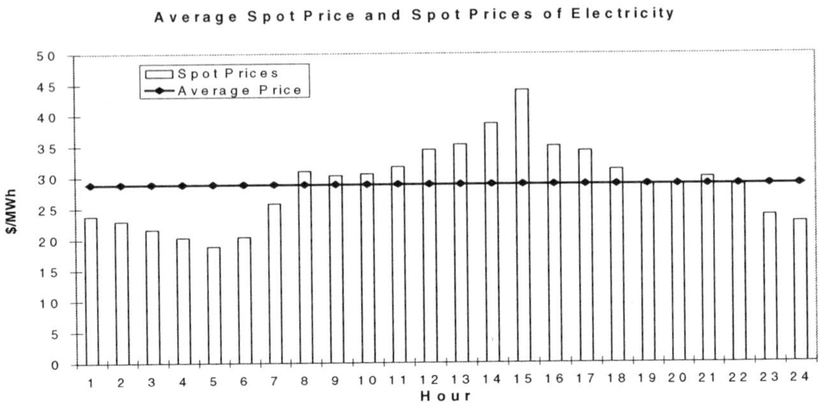

Figure 8-5Average versus actual spot price patterns

Chapter 9

Modeling Locational Price Differences

9.1 INTRODUCTION

In this chapter we will address the question of how market participants can quantify and hedge locational price risk. The work draws on results from the finance, economics and engineering community, attempting to find a middle ground that allows us to solve the unique problems facing the electricity industry. This includes developing methods for valuing newly emerging transmission dependent derivative contracts. Furthermore we examine the relationship of these new contracts with existing forward and option contracts on locational spot prices. We extend this analysis to include the valuation of investment opportunities in transmission assets, thus allowing a for-profit transmission provider to arrive at a market based valuation of a potential investment, based on observed forward and derivative prices.

9.2 LOCATIONAL PRICING AND MARKETS FOR TRANSMISSION

9.2.1 Modeling and Pricing Flows in Electric Power Networks

What differentiates electricity from other network based industries, such as telecommunications or transportation systems, are the complex physical constraints which govern the flow of power in transmission networks. Electricity cannot be sent point to point along a specified path. The flows are a nonlinear function of injections at the network nodes, and are governed by Kirchoff's current and voltage laws. In the regulated industry, utilities applied an optimal power flow (OPF) methodology, which combined power flow with the cost functions of their generation assets and arrived at least cost production schedules [47],[48]. With the creation of competitive power markets, the decision process of how much each generator should produce has moved from a centralized OPF problem to a decentralized auction scheme. Each producer decides how he wishes to bid his generation assets based on his own local objective function. The amount of power generated in each location then becomes a direct function of the market price of electricity in that area. The role of the independent system operator (ISO) is to determine a set of locational prices, such that the resulting injections of power by the generators do not violate the physical constraints of the transmission lines. Specifically, if the power flow from location A to location B in the network exceeds the capacity limit of the line, the system operator will want to raise the price at B and lower the price at A. This will presumably increase the local injection of power at node B, and decrease the injection at A, resulting in a decrease of the flow of power from A to B. This process is known as congestion management. Mapping this approach to a large scale network with non-linear relationships between flows and injections is a challenging task. There are currently two main congestion management methods commonly used in power systems: nodal and zonal pricing methods [49],[52],[53]. A third method, proposed in[55] and [54], is a generalization of the zonal pricing method and is based on congestion clusters.

9.2.2 Contracts for Transmission

Depending on the region, a number of different contracts on transmission are available to market participants to hedge their locational price risk. We will not describe each contract type in depth, but rather focus on two critical components which differentiate these contracts from one another. First we differentiate between physical transmission contracts, which assign the right to physically use the transmission grid, and financial contracts that result in a cash flow dependent on locational prices. Next we differentiate between contracts where the holder is obligated to use the transmission capacity (or receive or pay the difference between the spot prices at each location), and the case in which the holder has the option to use the capacity. It will be demonstrated that the difference in optionality has a considerable effect on the value of the transmission right. From here on we will refer to the types of contracts with no optionality as fixed transmission rights, while contracts with optimality are referred to as flexible transmission rights. How existing transmission contracts fall with respect to these criteria is outlined in table 1.

Type of contract	**Contract execution obligation**	
	Flexible (optional)	**Firm (mandatory)**
Physical	PTR, Ownership of transmission line	
Financial	Locational spread option	FTR

Table 9-1: Properties of Transmission Contracts

9.2.3 Valuing Transmission Rights

As transmission rights are being issued, either through auctions, or bilateral secondary markets, there is little consensus on how these rights should be valued. Consider the following possible approaches to valuing a transmission contract.

1. A supplier about to enter into a bilateral contract with a consumer, may value a firm transmission right in the following way. If the price specified by the bilateral contract is larger than the cost of the transmission right, combined with the cost of generating the power, then the supplier should accept the bilateral contract, and purchase the transmission right. This is a naïve valuation method, since it assumes that the supplier has to either purchase the transmission right or withdraw from the bilateral agreement with the consumer.

2. The spot market for electricity provides generators with a second option to supply their customers. In addition to the previous example where suppliers have to purchase the right to transmit the power from their plant location to the customer, they can sell the power locally on the spot market, and purchase power from the spot market at the customer location. This suggests that market participants ought to value transmission rights by projecting future spot price differentials and determine which delivery method is the most economical.
3. Assuming that an energy service provider always has the option to supply its load through the spot market, the value of a transmission right, either physical or financial, can be viewed as a derivative of locational spot prices. By switching the paradigm of the discussion into the field of financial derivates, we can start to address a number of new questions. What is the relative value of a transmission right contract, and of a set of forward contracts at the respective end points? How is the stochastic evolution of transmission contract prices related to changes in forward market prices? Is it possible to replicate a transmission right with a portfolio of forward and options contracts?

In this paper we will attempt to answer these questions by imposing dynamic constraints on the interaction of physical and financial processes on the network. A key assumption in this approach is that financial and physical rights are interchangeable in the network. In the following section we investigate this assumption in terms of the risk exposure of market participants.

9.2.3.1 Physical vs. Financial Risk

Participants in competitive power markets face two types of uncertainty: physical and financial. Physical risk refers to the possibility of having service to a customer interrupted. Financial risk is derived from uncertainty about future prices, and refers to the resulting variance in the market participant's profit.

In markets for transmission, the line between physical and financial rights is necessarily blurred. To better understand why this is true, consider the following example, where our world consists of two electricity markets (or two zones of the same market) connected by a transmission line. 1 supplier in market 1 has entered into a contract to supply physical power to a consumer in market 2. To enable this transaction, the supplier has also purchased a physical transmission right from 1 to 2. The transmission right, however, may be curtailed by the system operator under certain circumstances, such as a physical failure of the transmission line. Would such a curtailment represent a physical or financial risk to the parties in the bilateral contract?

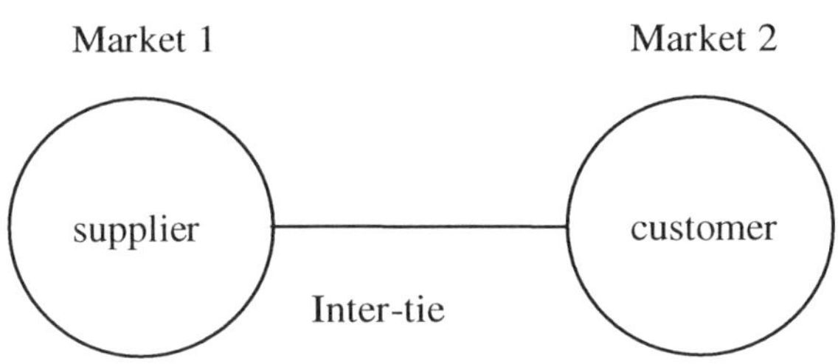

Figure 9-1 Example of two markets

A curtailment of the transmission right does not absolve the supplier from his obligation to serve the customer, but it makes it impossible for him to transmit his locally generated power to the load. This problem can be circumvented if the supplier sells his power in market 1 and purchases an equivalent amount of power in market 2. The physical risk of curtailment has then been transformed into a financial risk from the price spread between the two markets. There is also an inherent assumption that there is power available to be bought in market 2 at any price. This may not be the case, especially if the region is heavily dependent on imports coming through the downed transmission line. If power is not available, the load in the bilateral contract cannot be served and the transmission right curtailment represents a physical risk.

In this paper we will address primarily the question of how to manage financial locational risk in electricity markets. The assumption is that there is a secondary market for reliability, which addresses the problem of having sufficient reserves available to deal with contingencies. The authors recognize that in the current state of power markets in the United States, the question of reliability remains a major unresolved issue. Another motivation for considering the problem from a financial perspective is that even a physical failure to deliver is generally associated with a financial penalty. We can view this penalty as a default spot price, allowing us to transform physical risk into financial risk, as seen from the suppliers' viewpoint.

9.3 MODELING TRANSMISSION RIGHTS AS A DERIVATIVE ON SPOT PRICES

In this paper we will attempt to value three different types of assets: fixed transmission rights, flexible transmission rights and the ownership of a transmission line. Each of these assets can be thought of as a derivative of the locational spot price at the end nodes. Consider the setup described in figure(). We proceed to calculate the value of each asset at maturity, i.e. at the actual time of use of the transmission asset, financially or physically. We find that in each case, this value is a function of the underlying spread between the spot prices S^1 and S^2. In each case it is assumed that the transmission right held is in the direction 2 to 1, and the quantity is q MW.

The value of the fixed transmission right is simply the difference in the locational spot values at the time of maturity, and can thus be positive or negative:

$$V^{fixed} = q(S^1 - S^2).$$

The owner of the flexible transmission right will only exercise the contract if the price differential is positive. The payoff is therefore given by

$$V^{flex} = \max\{0, q(S^1 - S^2)\}.$$

The owner of the transmission line is assumed to collect all congestion rents, allowing him to profit irrespective of transmission line ownership:

$$V^{own} = q\left|S^1 - S^2\right|.$$

Identifying these relationships allows us to start to address the issue of the relative values of the above assets. Specifically we see that the payoff functions are linearly dependent, so that any two can be used to replicate the third perfectly. For example, by selling a fixed transmission right and purchasing two flexible transmission rights, one can guarantee the exact same payoff as from the physical ownership of the transmission line, independent of the price spread at maturity.

The analysis above focused only on the value of the transmission rights at maturity. In order to use these contracts as a part of hedging or speculation portfolios, however, the investor needs to be able to project the future values of the contract, and understand the dynamics of contract prices over time. To address this issue we need to postulate stochastic models for the underlying spot prices. We begin by reviewing the current state of the art modeling techniques.

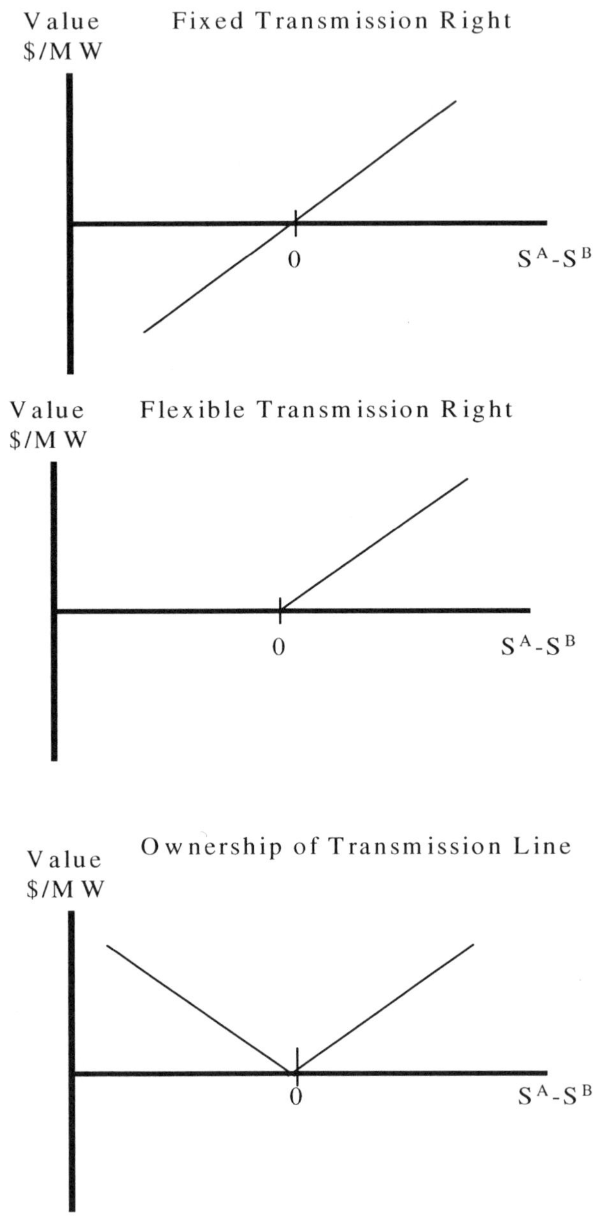
Value
$/MW
Fixed Transmission Right
0
S^A-S^B
Value
$/MW
Flexible Transmission Right
0
S^A-S^B
Value
$/MW
Ownership of Transmission Line
0
S^A-S^B

Figure 9-2

9.4 OVERVIEW OF EXISTING PRICE MODELS

The set of models used in pricing derivatives and managing financial risk are commonly referred to as volatility models. The purpose of these models are twofold: to characterize the probability distribution of future spot prices, and to estimate the correlation between future prices at different points in time. The most well-known application of volatility modes is the Black-Sholes option valuation formula, originally derived to value derivatives on equity. The basis for Black-Sholes is the assumption that the price of a stock, S, can be characterized by a random walk process known as Geometric Brownian Motion (GBM),

$$dS = \mu S dt + \sigma S dz \ ,$$

where z(t) is a continuous time Wiener process. For commodity markets, a variation of GBM is often used, based on the Ornstein-Uhlenbeck stochastic process,

$$dS = \kappa(\mu - \ln S) S dt + \sigma dz \ .$$

This process, also known as the mean reverting process, captures a property of commodity prices in which they tend to revert back to equilibrium levels after temporary shocks.

The models presented above can be extended to address derivative on the spread between two markets. The common approach is to model the spot price in each market as a separate Ito process. In the case of commodities, each spot price S^i is characterized by a mean-reverting process. The interaction between the markets is characterized by the correlation ρ between the Wiener processes driving each spot process.

$$dS^1 = \kappa^1(\mu^1 - \ln S^1) S^1 dt + \sigma^1 dz^1$$
$$dS^2 = \kappa^2(\mu^2 - \ln S^2) S^2 dt + \sigma^2 dz^2$$

$$dz^1 dz^2 = \rho dt$$

This approach was taken by Deng, Johnson and Sogomonian [44] in order to price locational spread options in electricity markets.

A number of questions arise, however, in the implementation of this modeling approach in electricity markets. Can the interaction between electricity prices at different locations really be captured through a simple correlation parameter? Will the correlation between markets remain constant over time, and if not, how does it vary in response to changes in market structure, expansion of the transmission grid, addition of new capacity, and growth in the demand? In this paper we will take a closer look at the dynamics of the uncertainties, which drive locational price differences. To do this we introduce a new type of volatility model, the bid-based model. The main deviation from the above-mentioned approaches is that we model price as an output form a process with distinct supply and demand states. As will be shown, this allows for a link between the physical and financial processes in the marketplace, and a better understanding of what determines the correlation between locational prices.

9.5 INTERACTIONS BETWEEN NEIGHBORING MARKETS

We use the Bid-based Stochastic Model to simulate the behavior of prices in a multi-market scenario. We estimate the value of a transmission right between two spot markets by modeling a joint evolution of loads and electricity prices in the adjacent markets, connected by a tie-line of fixed maximum capacity F_{max}^{12} [18],[19].

Scheduled transmission flows occur when there is cross-bidding between markets; that is, when loads or suppliers in one market decide that they are better off purchasing or selling their power in the neighboring spot market. A positive flow from market i to market j can be caused by two types of actions:

- *Supply cross-bidding:* Suppliers in market i decide to bid their power into market j. This causes b^i to increase and b^j to decrease.
- *Load cross-bidding:* Loads in market j decide to bid their demand into market i. This causes L^j to decrease and L^i to increase.

The net effect on price of the two actions is equivalent. Without loss of generality we decide to interpret all flows as the effect of load cross-bidding. To incorporate this behavior into the model, we introduce a new variable q_d^i, representing the actual quantity bid into market i at time d. The variable L_d^i is interpreted as the native load of the market, which is physically located inside the market's borders. Price in market i is a function of the total load and supply bid into this market:

$$S_d^i = e^{a^i q^i + b^i}, \quad i = 1..2$$

The relationship between q and L for the two-market example can be written as

$$\begin{aligned} q_d^1 &= L_d^1 + F_d^{12} \\ q_d^2 &= L_d^2 - F_d^{12} \end{aligned}$$

where F_d^{12} is the flow from market 1 to market 2, which can be positive or negative. The power has to be balanced between the markets

$$q_d^1 + q_d^2 = L_d^1 + L_d^2$$

and the tie-line flow F_d^{12} is bounded by F_d^{max}.

$$\left|F_d^{12}\right| = \left|q_d^1 - L_d^1\right| = \left|L_d^2 - q_d^2\right| \le F^{max}$$

The quantity of load, which cross-bids into the neighboring market, is calculated by assuming that market agents are rational. If there exists a price differential between the markets, the load in the expensive market will submit bids into the cheaper market. The magnitude of the cross bidding is limited by the capacity of the transmission line. Thus load bids will keep shifting from the expensive to the cheap market until one of the following occurs:

1. The prices equalize, thus removing any incentive for further cross-bidding.
2. The transmission line becomes congested, preventing the native loads from being supplied from the other market beyond a certain level.

The first case corresponds to the following mathematical condition:

$$S_d^1 = S_d^2$$
$$a^1 q_d^1 + b_d^1 = a^2 q_d^2 + b_d^2$$

The flow necessary to reach price equality $\hat{F}_d^{12}$, as a function of native load and supply states, is given by

$$\hat{F}_d^{12} = \frac{1}{a^1 + a^2}\left[\left(a^2 L_d^2 + b_d^2\right) - \left(a^1 L_d^1 + b_d^1\right)\right].$$

The transmission constraint F^{max} limits the flow both ways.

$$-F^{\max} \le F_d^{12} \le F^{\max}$$

The actual flow between the markets F_d^{12}, accounting for the limits, can therefore be written as

$$F_d^{12} = \max\left\{\min\left(\hat{F}_d^{12}, F^{\max}\right), -F^{\max}\right\}.$$

The prices in two markets, S_d^1 and S_d^2, are always equal, until the transmission flow reaches the maximum capacity. At this point prices will diverge, and the dynamics of the two markets will decouple.

9.6 VALUING A TRANSMISSION RIGHT

A flexible transmission right of x MW between market 1 and market 2 is interpreted as the right, but not the obligation, to transmit up to x MW of power from market 1 to market 2 on any day d. Furthermore we assume the transmission right to be firm, so it can not be curtailed under any circumstances. The expected daily profit from owning a flexible transmission right can therefore be expressed as:

$$C_d^{12} = E\left(\max_d \left(P_d^2 - P_d^1, 0 \right) \right)$$

The value of a transmission right between markets 1 and 2 thus is equivalent to the value of a spread option between the two markets.

9.7 SIMULATION BASED VALUATION

Tab. 3: BSM parameters, used in simulations; equal in both markets

	α	κ	σ_m	σ_m^{δ}	a
Load	0.3	50	500	2000	0.00015
Supply	0.3	0.05	0.0005	0.25	

Tab. 4: Monthly BSM parameters, used in simulations

	μ_m	
	Market 1	Market 2
Load	13000	20000
Supply	1.7	1.7

We estimate the value (in \$/MWh) of owning a flexible transmission right for one day, thirty days from today. To better illustrate the qualitative effects of moving to a multi-market environment, we have ignored the dynamics of the mean-process δ in the single-market bid-based price model, forcing long term means of both processes to be zero, $\delta^{Li} = \delta^{bi} = 0$, $i = 1,2$ for all times. The parameters and initial states for the demand and supply processes in both markets are given in tables 2 and 3 respectively. It should also be noted that, for this particular example, we assume to random walk processes driving the load and supply processes in each region to be uncorrelated.

We simulated the behavior of the two market model for various values of the maximum transmission capacity, varying from 0 to 3500. For each scenario, the model was then run 10,000 times for a 31-day period. The plots show the correlation between locational prices and the value of the flexible transmission right on the 31st day.

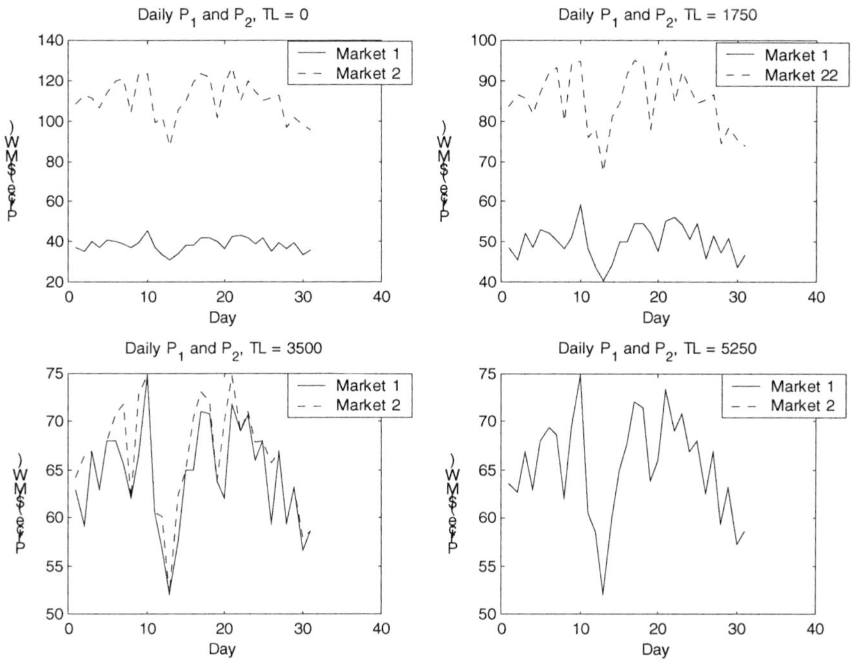

Figure 9-3: Simulated joint spot price dynamics for two markets joined by a transmission line of varying capacity.

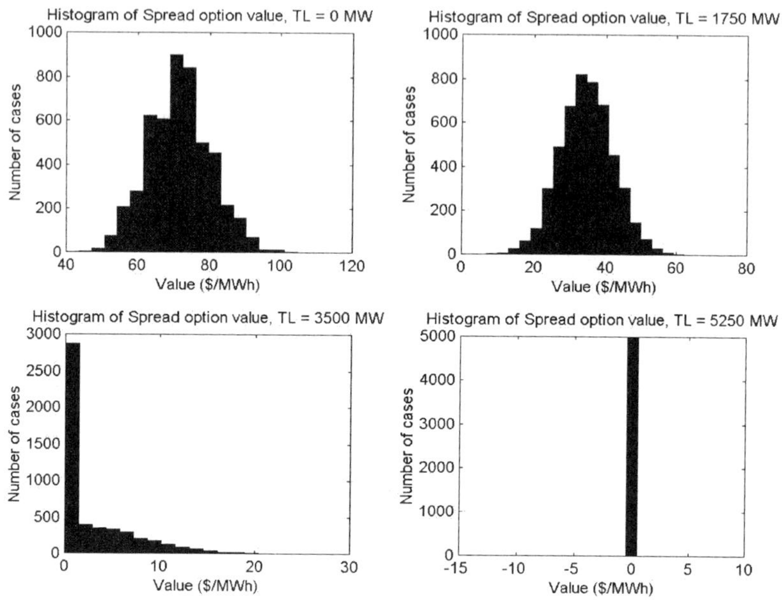

Figure 9-4

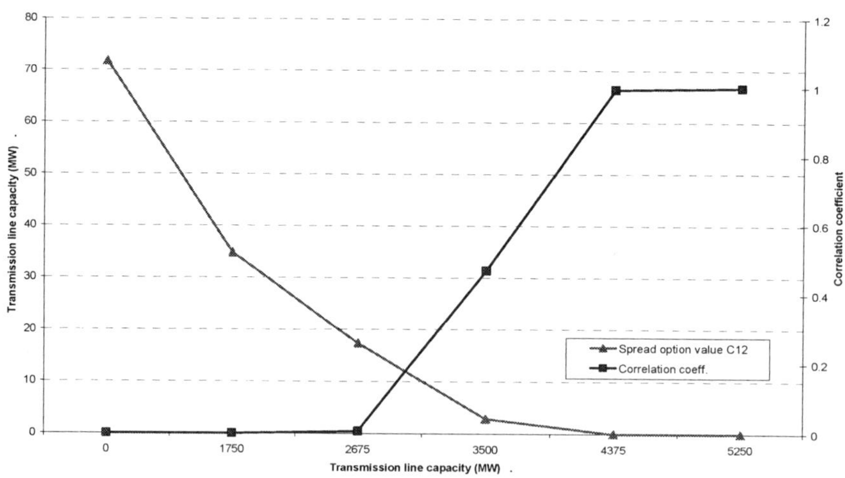

Figure 9-5 Spread option value C_{12} and price correlation coefficient in various TL capacities

As the size of the transmission line increases, the correlation between the locational prices increases. Consequently the probability of the prices diverging is reduced, so that the value of the spread option, and the transmission right, decreases.

9.8 DYNAMIC HEDGING

A simulation based approach allows the user to estimate the expected cash flow from a contract, and the associated risk, or variance, of the cash flow. In addition, traders also need to understand the relationship between the contract price and the current value of the underlying state variables. This knowledge allows the user to cancel out risk between a variety of contracts, as long as the risk is derived from a limited number of underlying sources. A well-known example is the case of delta hedging. As shown in section 9-4, the Black Scholes model assumes that all uncertainty affecting the price of a stock can be modeled as originating from a single Wiener process. By Ito's lemma, all derivatives of the stock price will follow Ito processes driven by the same Wiener process as the stock price. This result allows the trader to hold a combination of the derivative and underlying stock so that the uncertainty cancels itself out exactly. The ratio of the stock (S) to the derivative (g) required to eliminate the uncertainty is known as the delta of the derivative, given by

$$\Delta_g = \left.\frac{\partial g}{\partial S}\right|_S .$$

The delta hedge will only cancel out the uncertainty for a given stock price S. Since the stock price evolves continually, the value of delta will also constantly change. To perfectly eliminate risk, a trader therefore needs to rebalance his portfolio constantly. The process of rebalancing in response to changes in the underlying asset price is known as dynamic hedging. The effectiveness of a dynamic hedging strategy depends in large part on the accuracy of the underlying model. Furthermore, due to transaction costs and other real market constraints, a continuous replication strategy cannot be implemented. Instead, traders must rely on approximate strategies where the portfolio is rebalanced at discrete time steps. In this case, one must analyze the robustness of the linearization, as spot prices diverge from the initial operating point.

9.8.1 Dynamic Hedging and the Bid-based Model.

In the case of electricity, uncertainty in the spot price process cannot be well described through a single random input. As discussed in the presentation of the bid-based model, there are a number of factors influencing the bid behavior of suppliers and consumers. The model describes four sources of uncertainty, a long and short term process for demand and supply respectively. In the case of two markets linked by a transmission line, the joint dynamics is governed by four state variables (demand and supply states in each area), with eight associated random inputs. This approach may seem overly detailed. Why not apply a reduced order model of the regional price, as proposed by Deng et al. [44]. The reason lies in the asymmetrical effect of the transmission line flow on the distribution of the price spread between the markets. To illustrate this effect, consider the two market environment as described in the simulated example in section 6-5, with a transmission line of capacity 3,500MW. We are interested in the sensitivity of the price spread to a change in one of the underlying states, L or b. Figure 9-6 shows a plot of the price spread as a function of L_1 as the other three states (L_2, b_1, and b_2) are kept constant at the initial values given in table 6-3. The figure illustrates an interesting property of the sensitivity of the price spread to demand changes in region 1. The plot separates into three distinct regions. In the middle region, the transmission line is uncongested, and the price spread therefore is zero. In this case, the two areas act as a single market with a single price. We refer to this region as the 'dead band' since the price spread is insensitive to changes in the underlying states. The left region represents the case when the transmission line is congested in the direction of 2→1. In this case the markets decouple. The flow on the transmission line is fixed at its maximum value, so that changes in the state variables of one market do not get transmitted through changes in the flow to the neighboring market. The uncertainty in the spot price in each region is governed by the uncertainty in the local load and supply variables. However, the price spread (S_1-S_2) is always negative. Conversely, the right hand region of the plot shows the behavior of the price spread when the line is congested from market 1 to 2. In this case the price spread is strictly positive. A similar type of behavior can be observed when plotting price as a function of the supply state.

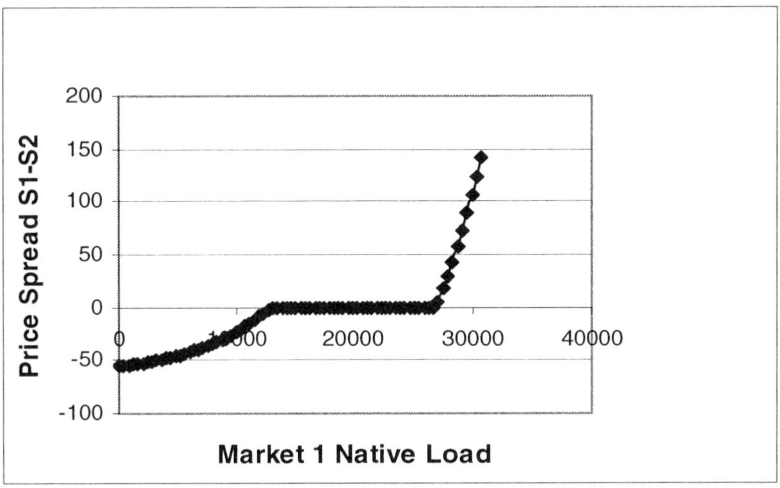

Figure 9-6: Price spread between markets 1 and 2 as a function of native load in market 1.

9.8.2 Implications on the Valuation of a Spread Option

An important implication of the property displayed in Figure 9-6 is the asymmetry of the distribution of the price spread. Recall the price based approach to modeling spread options. This model implies that the future spot prices in each region are lognormally distributed random variables. As a result, the probability of the two prices being exactly equal is infinitesimal. Furthermore, if the prices were equal for a given instance, there would be a non-zero, though not necessarily equal, probability of the spread being positive or negative in the next instance. This is a qualitatively different behavior than what we would expect to see according to the bid based model. Consider again the case where we allow only the load in area 1 to vary, keeping all other states fixed. The stochastic process governing load dictates that future load levels are normally distributed random variables.

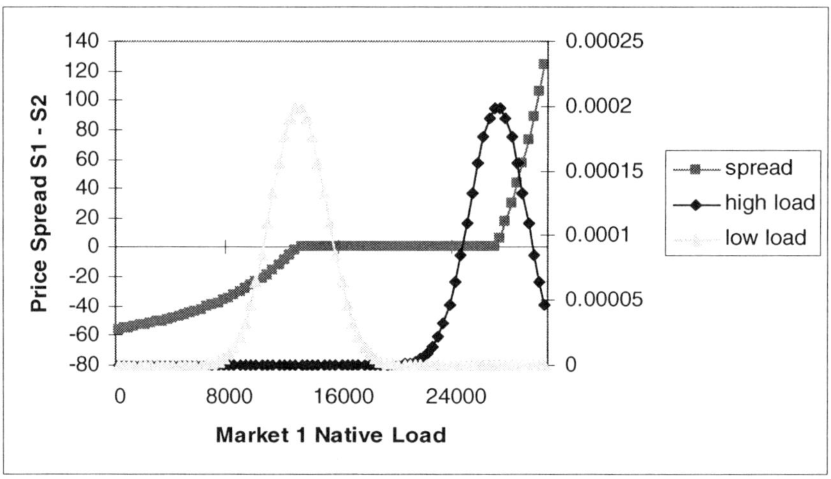

Figure 9-7: Price spread between two markets as a function of native load in market 1, with superimposed load distributions.

Figure 9-7 shows two normal distributions superimposed on the plot of the sensitivity of price spread to changes in L_1. The two distributions, high load and low load, represent the projected distributions of load on a high demand and low demand respectively. The means of the distributions are 13,000 MW and 23,000MW, and the standard deviation in both cases is 2,000MW. The figure illustrates an interesting condition. Each load distribution covers two regions of the graph, but the probability of ending up in the third region is nearly zero. In other words, a load change during a high demand period could cause the transmission line to move from uncongested to congested in the 1→2 direction. However, it is extremely unlikely that a load change would be large enough to cause congestion in the 2→1 direction. Conversely, during a low demand period, the transmission line will be either uncongested or congested in the 2→1 direction, but is unlikely to be congested in the 1→2 direction. Combining this result with the link between the direction of congestion and the sign of the price spread gives an interesting result outlined in the table below.

Expected Load Level (L^1)	Probability $S^1>S^2$	Probability $S^2>S^1$	Probability $S^1=S^2$
15,000 MW	$2*10^{-15}$	.5	.5
25,000 MW	.5	$3*10^{-12}$	.5

If one attempt to find a set of parameter for the traditional spot price model, which produces a similar set of probabilities for the price spread, one would find this impossible. The bid based dictates a qualitatively different behavior of the price spread.

9.8.3 Replicating a Flexible Transmission Right under the Bid-based Price Model

Closely related to dynamic hedging is the dynamic replication problem. In replicating a derivative, we create a portfolio of contracts which have the same payoff. Dynamic replication is the practice of creating a dynamically changing portfolio which matches the value of a derivative over time. Dynamic replication has a significant impact on the pricing of derivatives contracts. If a derivative can be perfectly replicated, then one can create a portfolio with exactly zero payoff by purchasing the derivative and selling the replicating portfolio. If the price of the replicating portfolio differs from that of the derivative, this represents an arbitrage opportunity.

In this section we attempt to replicate a flexible transmission right, or equivalently a locational spread option. The option is purchased at time t, and matures at time T. The price of the option is $C^{12}(t,T)$, and its payoff at maturity is given by

$$payoff = \max(0, S_T^1 - S_T^2)$$

The nature of the price spread lends itself nicely to this problem. Consider three states of the transmission line.

1. State 1. The line is congested in the direction 1→2. In this case $S^2 > S^1$.
2. State 2. The line is uncongested. In this case $S^1 = S^2$.
3. State 3. The line is congested in the direction 2→1. In this case $S^1 > S^2$.

Next consider the following set of replicating portfolios.

1. Portfolio 1. Purchase a forward contract in market 1, sell forward contract in area 2.
2. Portfolio 2. Do nothing.

Now consider the payoff at maturity from the call option, as well as the two replicating portfolios, under each of the three transmission line states, (assuming the forward contracts are prepaid).

	State 1	State 2	State 3
Call option	$Max(0, S^1-S^2)=0$	$Max(0, S^1-S^2)=0$	$Max(0, S^1-S^2)= S^1-S^2>0$
Portfolio 1	$S^1-S^2<0$	$S^1-S^2=0$	$S^1-S^2>0$
Portfolio 2	0	0	0

The table shows that the call option has the same cash flow as portfolio 1 in states 2 and 3. The option has the same cash flow as portfolio 2 in states 1 and 2. Combining this with the probability distributions displayed in Figure 9-7, we find that in many cases a very simple replication strategy will do. For the high demand period, the transmission line is very likely to be in state 2 or 3, so portfolio 1 provides a good replication. For the low demand period, the transmission line is likely to be in state 1 or 2, and therefore the payoff from the spread option is zero and the replicating portfolio is empty.

9.9 GENERALIZATION OF THE MODEL TO A 3 NODE EXAMPLE

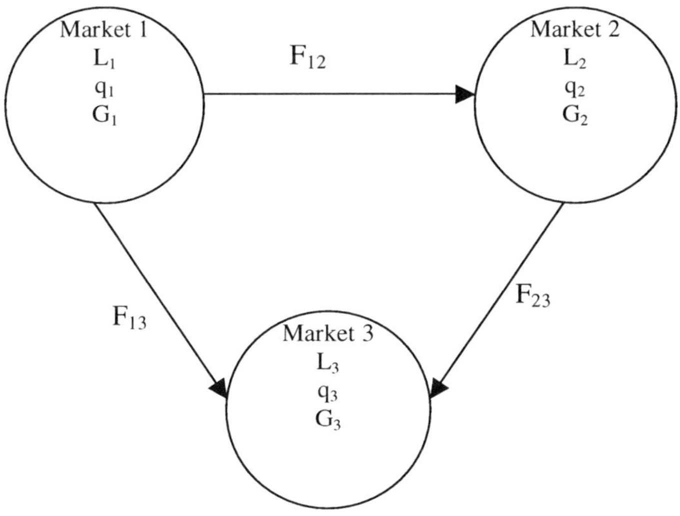

Figure 9-8 A three market example

Power flows in a three-market example are in contrast to the two-market example calculated using optimal power flow, based only on the flow of active power between generators and loads (DC-OPF). Generation in each of the markets is dispatched in a non-linear manner according to a composite generation cost function using constrained optimization. As in the two-market example, we interpret all flows as the effect of load cross-bidding, denoting L_d^i as the native load of each market and q_d^i as the actual quantity bid into market i at time d. The price in market i is a function of the total load and supply bid into this market. The relationship between q and L can be expressed as

$$q_d^i = L_d^i + \sum_{j=1}^{3} F_d^{ij}, \quad j \neq i, \quad i = 1..3$$

where F_d^{ij} is the flow from market i to market j. Due to conservation of energy, the total amount of generation in the system matches the total native load,

$$\sum_{i=1}^{3} G_d^i = \sum_{j=1}^{3} L_d^j ,$$

while the power balance equation postulates that the amount of native load equals the amount of bid-in load.

$$\sum_{j=1}^{3} L_d^j = \sum_{k=1}^{3} q_d^k$$

In the case of a perfectly constrained network, flows on transmission lines equal $F_d^{ij} = 0, \quad G_d^i = L_d^i, \quad i = 1..3$. Otherwise, the total generation of the system is allocated to the individual markets according to the minimum cost criterion, with a cost function J:

$$J = \sum_{i=1}^{3} \left(a_i \left(G_d^i - L_d^i\right) + b_i\right) = \sum_{i=1}^{3} \left(a_i q_d^i + b_i\right).$$

Constrained optimization takes into account the transmission limits of the lines F_{max}^{ij}.

9.9.1.1 Simulation of the Three-Market Model

To investigate the value of a transmission right in a three-market setting (in \$/MWh), we have again looked at the time frame of one day, thirty days from today. The dynamics of all three markets have been modeled according to the bid-based stochastic price model. The respective parameters for the load and supply process in all three markets were identical, except for the mean-value of the load in market 3, μ_L^3. To create a price differential, μ_L^3 was set to 14,000 MW, 1,000 MW higher than in the neighboring markets.

$$\mu_L^1 = \mu_L^2 = 13.000 MW$$
$$\mu_L^3 = 14.000 MW$$

The algorithm for calculation of native generation and prices in each of the three markets for a 31-day period is shown in Figure 9-9. Using the respective daily loads, the DC OPF is used to compute native generation G^i, i = 1..3, according to the cost function J and transmission constraints F_{max}^{ij}. From native generation G^i, load L^i and supply curve shift b^i, market i price P^i is then calculated. This algorithm is ran in a loop where the transmission capacity F_{max}^{ij} of all three lines is gradually increased in 6 steps from 0 MW to 3,750 MW. The plots in show loads and prices for the period of 30 days.

As the transmission capacity is gradually increased the flows on the lines increase as well, as shown in Figure 9-10, yet the effects of congestion are eminent on the first four graphs. At the same time, correlation in prices among markets increases as the lines become less congested, see Figure 9-10.

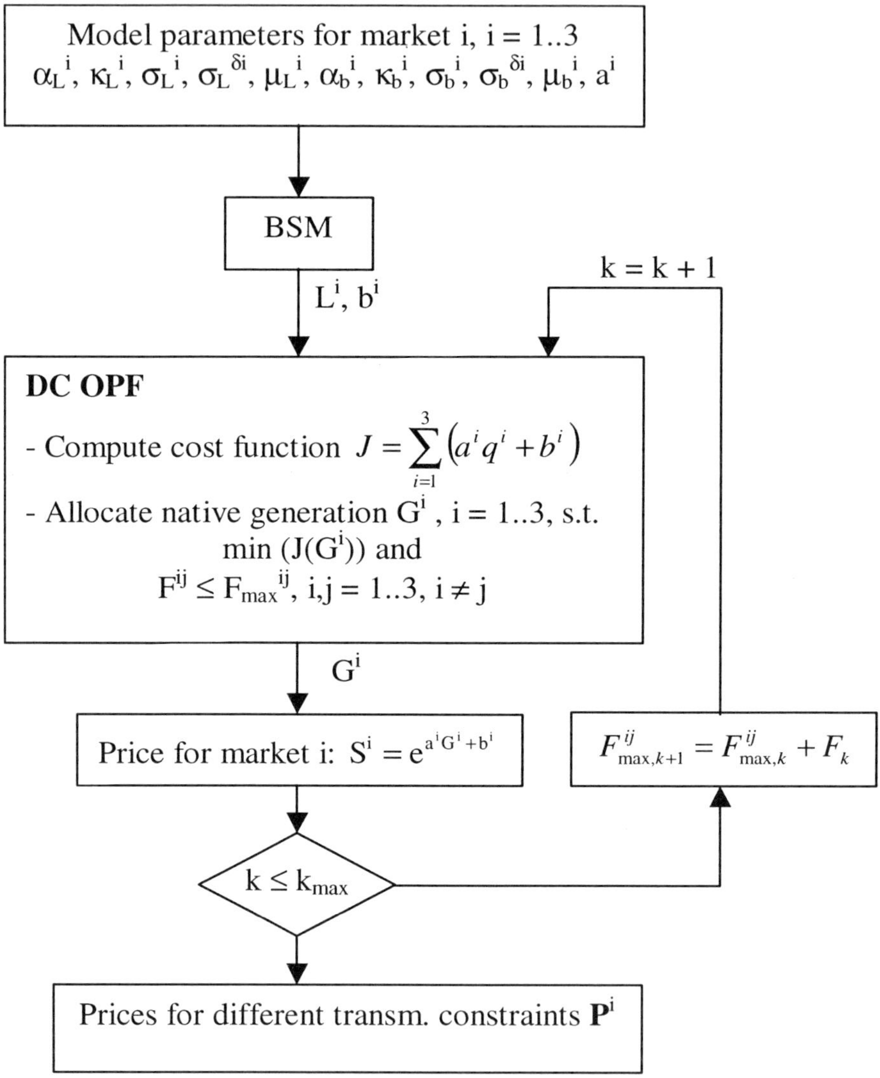

Model parameters for market i, i = 1..3
$\alpha_L{}^i, \kappa_L{}^i, \sigma_L{}^i, \sigma_L{}^{\delta i}, \mu_L{}^i, \alpha_b{}^i, \kappa_b{}^i, \sigma_b{}^i, \sigma_b{}^{\delta i}, \mu_b{}^i, a^i$
BSM
L^i, b^i
k = k + 1
DC OPF
- Compute cost function $J = \sum_{i=1}^{3} \left(a^i q^i + b^i\right)$
- Allocate native generation G^i, i = 1..3, s.t.
min $(J(G^i))$ and
$F^{ij} \leq F_{max}{}^{ij}$, i,j = 1..3, $i \neq j$
G^i
Price for market i: $S^i = e^{a^i G^i + b^i}$
$F^{ij}_{\max,k+1} = F^{ij}_{\max,k} + F_k$
$k \leq k_{max}$
Prices for different transm. constraints $\mathbf{P}^i$

Figure 9-9

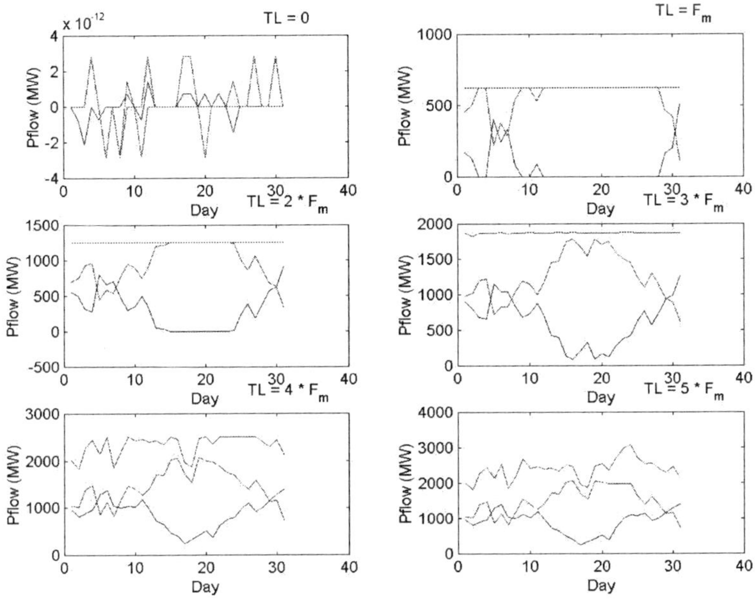

Figure 9-10 Daily flows in 3 markets with various transmission capacities

Chapter 10

Investment Dynamics and Long Term Price Trends in Competitive Electricity Markets

10.1 INTRODUCTION

Our attention is often drawn to competitive power markets in times of crisis. Electricity markets are now experiencing unprecedented levels of price volatility. Each time prices spike there are those who call for regulatory intervention to protect customers and keep large suppliers from exploiting shortages. Most recently, the power crisis in California resulted in both financial losses for the load serving utilities, and the physical loss of power for some of their customers. This resulted in a public reevaluation of the success of deregulation in lowering the cost of electricity for the consumer, as well as in the more basic premise of keeping the lights on.

This chapter addresses the question of price trends in competitive electricity markets, both in terms of economic efficiency and physical reliability. The key to successful deregulation does not lie in the daily operation of the system. Short term optimality is always easier to achieve in a regulated, centralized industry. The goal of deregulation should be to provide the right incentives for new investment, and the development of new innovative technologies. This evolution occurs on a longer time scale, and with its own dynamic constraints. In order to transition successfully to a deregulated environment, the regulators must recognize the nature of the decentralized decision process which governs investment. This includes modeling the effect of price signals on investment, and understanding the impact of delays in information on the price dynamics as well as on the physical power balance of the system. Only if these relationships are fully understood should a regulator attempt to intervene into the marketplace.

10.2 A LONG TERM MODEL FOR ELECTRICITY PRICES

The model characterizes spot price as a function of two state variables: L representing the total market load, and b representing the current state of supply. In this chapter we focus on the long term dynamics of the electricity prices. A more detailed discussion of the model presented in this section, which captures short term deviations in prices, can be found in [18]. The demand for electricity is assumed to be inelastic, while the basic shape of the aggregate supply curve is characterized by an exponential function, with a stochastic shift parameter b. The average price in a month m can then be written as

$$S_m = e^{aL_m + b_m}$$

where a is a fixed parameter characterizing the shape of the bid curve.

10.2.1 Stochastic Demand Process

Demand for a given month is modeled as the sum of a deterministic component μ, and a stochastic component δ.

$$L_m = \mu_m^L + \delta_m^L$$

where μ_m^L captures the seasonal behavior of the load. The state δ_m^L represents the long term uncertainty in load, which grows stochastically with drift κ and volatility σ.

$$\delta_{m+1}^L - \delta_m^L = \kappa^L + \sigma^L z_m^L .$$

10.3 MODELING INVESTMENT DYNAMICS

Having developed a model for the stochastic growth of demand in the market, one can now address the question of how new generation capacity is added to the system in response to the load growth. It is assumed that the decision process for investing generation assets is decentralized. Each investor makes decisions in order to maximize his own utility, and there is no higher level entity coordinating investment behavior in the marketplace. For an in-depth analysis of investment decisions under uncertainty, see [6]. The rate of investment will not be governed by projections of overall demand and supply mismatches, as was the case in the regulated industry. Instead, investors react to price signals from the market in making their decisions. While price signals are inherently linked to the supply and demand levels, this change from a physical to a financial investment signal has profound effects on the dynamics of investment, and ultimately on the physical reliability of the system.

10.3.1 Backward Looking Investment

In the first model, it is assumed that the investor observes a moving average of the last 12 months of spot prices, S^{ave}. He compares this value to the index, I, of the available technology to invest in. I reflects the marginal cost of running the new unit, as well as the installation cost. If the average spot price rises above the index value, one starts to observe new investment in the market. The greater the differential between S and I, the higher the rate of investment, this difference is referred to as the investment signal. The parameter G determines the rate of investment in response to an investment signal. G can be thought of as reflecting the availability of capital in the market. Finally negative investment, that is the removal of capacity from the system in response to low prices, is not allowed. As with the load model, the stochastic component d is separated from the seasonal component m,

$$b_m = \mu_m^b + \delta_m^b .$$

The stochastic component evolves according to the following dynamic equation,

$$\delta_{k+1}^b - \delta_k^b = \max(0, G(S_k^{ave} - I_k)) + \sigma^b z_k^b$$

where

$$S_k^{ave} = \frac{1}{12}\sum_{j=1}^{12} S_{k-j} .$$

The model is backward looking because the investment decision reflects the previous 12 months of spot prices. In a market where investment decisions are made based on historical spot prices, there is an inherent delay between increased spot price levels and increased investment. Due to this delay, investors will continue to inject capital after spot prices have declined below critical levels. In a market with growing demand, this results in cyclical swings of high and low spot price periods, as investors alternately overshoot and undershoot their optimal investment levels. This effect is lost in standard economic equilibrium models, where it is assumed that suppliers are able to immediately take advantage of price increases. Another critical element in investment dynamics, is the delay between the time that a decision to invest is made, and the time that the new generation plant is actually connected to the power grid. This delay has two components. The first is the time it takes for the plant to be licensed by the regulators. The system operator goes through an extensive study on the effects of each new plant on the network, and approval can take over a year. Next there is the production and installation time of the actual generator. Put together these delays can block the markets ability to correct for generation deficiencies, further accentuating the cyclical price behavior observed above. This delay is accounted for by introducing the parameter τ in the dynamic equations governing investment,

$$\delta_{k+1}^b - \delta_k^b = \max(0, G(S_{k-\tau}^{ave} - I_{k-\tau})) + \sigma^b z_k^b .$$

The longer the delay, the greater the tendency for extreme price spikes followed by periods of suppressed price levels. The interaction between spot price levels and the investment decision, including the delays, is depicted in figure 10-1.

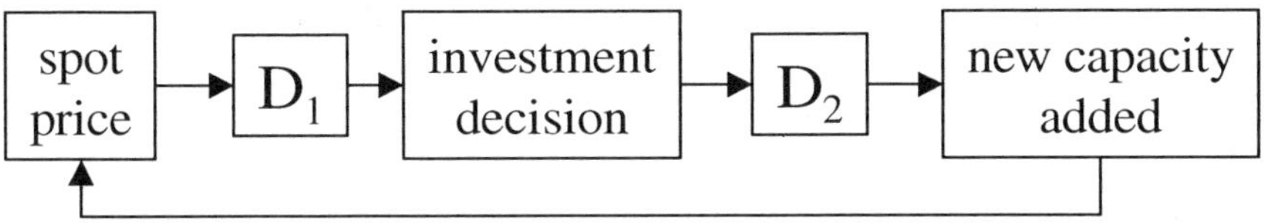

Figure 10-1

Figure 10-2 shows a simulated comparison over a 100 month period of market behavior without delays, and with a six month delay period. The parameters used in the simulation are provided in Table 10-1. It should be noted that these simulations are provided to gain a qualitative understanding of market behavior, rather than quantitative predictions.

Load	μ^L=12,000	σ^L=100	κ=100
Supply	μ^b=1.2	σ^b=.01	G=.003
other	$a=5*10^{-4}$	I=150	τ=0,6

Table 10-1

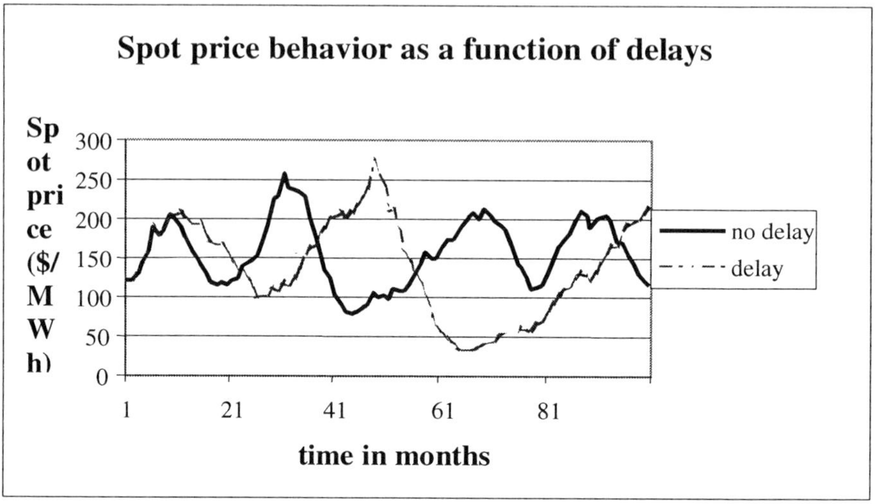

Figure 10-2

10.3.2 Forward Looking Investment

Two sources of delays in the investment dynamics have been identified: a delay from the price signal to the investment decision, and a delay from the investment decision to the installation of the plant. Both of these delays could be negated if investors were able to project future price trends. A long term price estimator would allow investors to base their decisions on projected future revenues, rather than historical data. This would have a stabilizing effect on the market, and eliminate much of the cyclical price behavior.

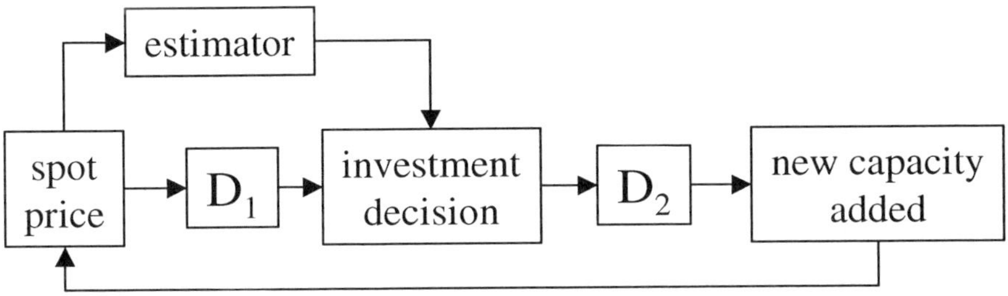

Figure 10-3

The challenge in the estimation problem lies in the fact that it requires the user to model the decision process of all other investors. The problem may be tractable in the case where there is sufficient historical data available to estimate the cumulative investment rate in response to market price (the G parameter in our model). However in the early stages of a market, such as the current situation in the United States, one would be forced to arrive at this parameter by deriving likely competitor strategies. This would be a very complex game theoretic problem, where the outcome would depend on how sophisticated market participants are in their decision process.

10.3.2.1 The Role of Futures Markets

In the context of forward looking investment, futures markets play an important role as an information provider. It is questionable whether futures prices truly reflect the expected value of future spot prices, but the prices do reflect information which is not present in historical spot prices. For instance, power marketers keep a close watch on the permit requests and manufacturing orders for new generators. This gives the marker an estimate of the amount of new generation capacity which is likely to be added in a given area in the near future. The information is incorporated into the marketer's futures trading strategy. If a region has a current generation shortage, and accordingly high spot prices, but there is an abundance of new turbines in the manufacturing or permitting stages, the futures prices will tend to be depressed. If investors observe the futures market, and if the new generation capacity in progress is accurately reflected in the futures prices, it will prevent over- and under-investment, thus stabilizing the spot price dynamics. There are two critical properties which futures markets must satisfy in order to govern investment dynamics effectively.

- **Liquidity:** The volume traded on futures markets is not necessarily proportional to the total load on the system. Instead it reflects market participants' desire to hedge their positions, or to speculate on future spot price levels. In order for the futures price to be a useful signal to investors, it has to be credible. That is, one must be able to buy and sell power in significant volume at or near the price quoted in the exchange. This in turn requires that there exist a large number of participants who actively trade in the market.

- **Duration:** The duration of a market refers to the longest time to maturity of all contracts currently trading in the market. If a market has a duration of 12 months, then a contract which matures April 1, 2002, will start to trade on April 1, 2001.To understand the importance of the forward market duration, consider the position of an investor who is contemplating financing a new power plant. The plant is estimated to take one year to be built and permitted. The investor is willing to undertake the project, if he expects to recover his initial capital investment in five years. To solve the decision problem, the investor must project the cash flow from the plant, and therefore the spot price levels, six years into the future. A futures market with a duration of a year or less has a limited value to an investor since he expects no cash flow until the plant is finished. A market with a duration of three or five years however, would allow the investor not only to make a market based estimate of future cash flows, but also lock in some of these revenues by selling futures contracts.

The two objectives, liquidity and duration, can be contradictory. By increasing the duration of the market, one increases the number of different contracts traded simultaneously, since each delivery month is a separate contract. This makes it more difficult to find two counter parties willing to trade at the same contract at the same time. Currently, the two main exchanges, NYMEX and CBOT, trade contracts up to fifteen months prior to delivery. This is not a sufficient time horizon for an investor seeking to value or hedge a new plant. At the same time, the exchanges are experiencing a lack of liquidity even for near term contracts.

We will not attempt to simulate the impact of forward markets here, since it would require us to make unfounded assumptions about traders' strategies. There are a few points which need to be studied carefully as more data becomes available from the futures exchanges.

1. To what extent do forward prices contain information which cannot be derived from historical spot prices?
2. Do investors depend heavily on futures price signals in making their investment decisions, or do they tend to wait until price changes appear in the spot markets?
3. Does the presence of a liquid futures market have a stabilizing effect on the spot market, eliminating periods of extreme over and under capacity?

10.4 A DYNAMIC NOTION OF RELIABILITY

When capital investment fails to keep up with load growth, there are two measurable effects in the market. The first is an increase in the spot price, as discussed in the previous section. The second effect is a reduction in the available generation reserve R, defined as the amount of unused generation available in the market as a fraction of the total load,

$$R_k = \frac{C_k - L_k}{L_k},$$

where C is the total capacity of all available generation assets. In a market with little or no demand elasticity, retaining a generation reserve is the only means of avoiding customer curtailments or blackouts as a result of unexpected load spikes or generation outages. The Federal Energy Regulatory Committee (FERC) sets guidelines for how much generation reserve each region should retain. It is the job of the independent system operators (ISOs) to enforce these reserve requirements. The system operator will do this by contracting generators to be in a stand-by mode. The compensation paid to these generators is determined through auctions similar to the electricity spot market. The problem is that if there is not enough total generation capacity in the market, the ISO will be unable to purchase reserve generation at any price. Furthermore, the ISO is not allowed to build or own generation assets. The system operator is therefore unable to guarantee that the system will meet the reserve margin. The reliability of the system can only be ensured by the addition of new generators, and investment in these plants is determined by for profit market participants. The reliability of competitive electricity markets is therefore directly coupled to the spot market price dynamics.

To illustrate the link between reliability and spot price dynamics, the model is further amended. Starting with a total capacity equal to the initial load, plus a reserve margin X,

$$C_0 = (1 + X)L_0 .$$

Every time there is new investment in generation, reflected in the supply state δ^b, there is an associated increase in the total available capacity C,

$$C_k = C_0 + \frac{1}{a}\left(\delta_k^b - \delta_0^b\right),$$

recognizing that a 100MW increase in L is perfectly offset by a (1/a)*100MW increase in b.

10.5 EFFECTS OF GOVERNMENT POLICY

In periods of high price levels, consumer advocates can put pressure on the government to impose price caps on the market. The argument is that suppliers are taking advantage of the generation shortage in order to drive up prices, either by withholding their generation or bidding it in at inflated price levels. The issue of 'fair' pricing of electricity will not be addressed here. Instead, we will try to answer the question of whether price caps are an effective means of reducing price levels in the long term. To do this the market is simulated under two conditions. The first is without a price cap, as shown above. In the second case, a price cap is introduced, leading to the condition

$$S_k = \min(cap, e^{aL_k - b_k})$$

where 'cap' is the \$/MWh price cap imposed by the regulator.

From the simulation it is clear that while the cap eliminates periods of high prices, it also raises price levels during the low price cycles. This result is easy to understand if one goes back and examines the signal which drives new investment:

$$S_k^{ave} - I \, .$$

By reducing price levels when supply is scarce, the regulator reduces the rate of new investment into generation. As a result, prices drop off at a slower rate, causing higher future spot prices. In the case described, the average power price is higher in the case where price caps are imposed .

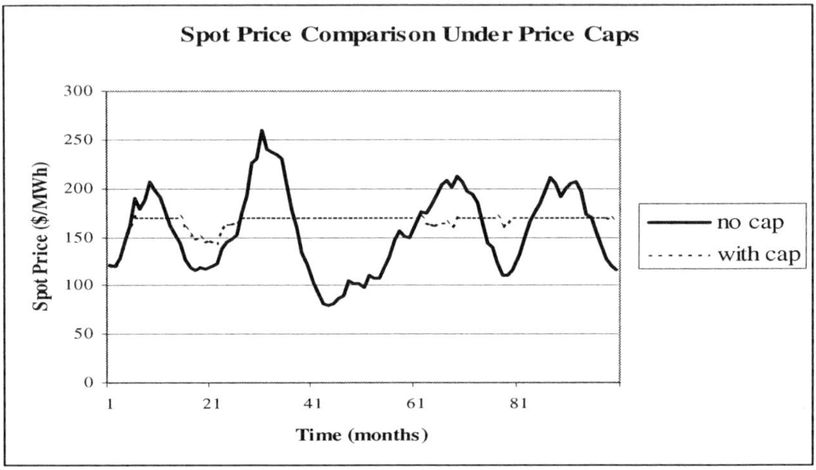

Figure 10-4

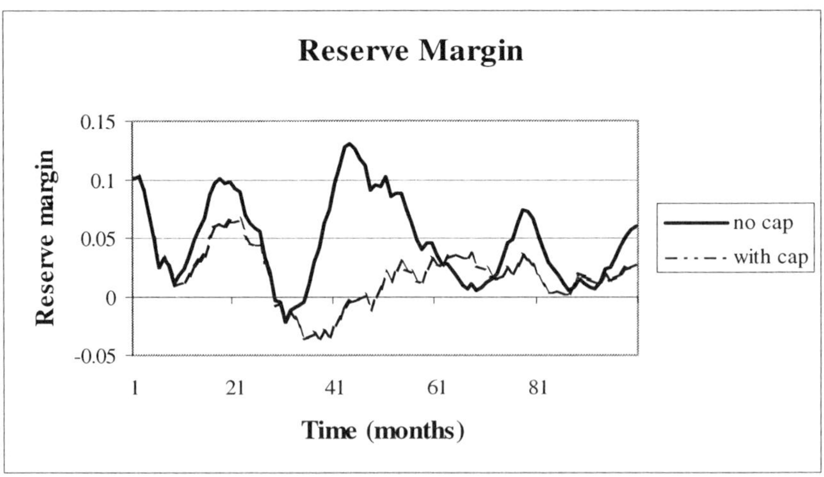

Figure 10-5

10.5.1 Comments on Simulation Results

The simulation demonstrates that reserve market levels tend to be lower in cases where a price cap is imposed by the regulators. When the price caps are near the critical investment level I, one starts to observe negative reserve levels. In these instances the system operator must order the curtailment of some customers, possibly through rolling blackouts, in order to prevent the collapse of the entire system. The simulation illustrates a trap which the regulator must avoid. By imposing price caps, the regulator succeeds in eliminating price spikes from the market. At the same time, the investment rate starts to drop off, thereby increasing price levels on the low end of the price cycle. The net effect is a flattening out of the price trend, which may actually raise the average price of electricity over a multiyear period. At this point it is tempting for the regulator to force down the average price by further reducing the level of the price cap. ***This is a dangerous move, because it decouples the spot price level from the economic reality of supply and demand.*** The scarcity of supply is not translated into high prices, and therefore the economic signal to investors to build new plants is blocked. Eventually the physics will catch up with the economics, as the available generation will no longer be able to meet demand, resulting in curtailments. The critical price level ($S^{critical}$), at which point investment will no longer keep up with demand growth, is given by

$$G(S^{critical} - I) = a\kappa .$$

If the price cap is set below $S^{critical}$, then investment cannot keep up with load growth, and the system is invariably headed towards blackouts.

The implications of the results in this paper must not be interpreted as rejecting all forms of regulatory intervention in general, and price caps in particular. There may be instances where it is necessary for the government to set temporary limits to the price in a market to prohibit suppliers from exploiting shortages. What the model illustrates is that the regulator must be very careful in setting these limits. Price caps must be set higher rather than lower, to ensure that the economic feedback is not blocked, and that market forces are allowed to bring the system back to stable price levels. Once price caps have been put in place at a too low level, they become increasingly difficult to remove as the generation shortage worsens.

10.6 CONCLUDING REMARKS

This chapter addresses the interplay between spot price levels and investment into new generation capacity in competitive electricity markets. The problem was addressed from the viewpoint of economic efficiency as well as the physical reliability of the system. Special emphasis was placed on the dynamic properties of the investment process. It was shown that delays caused by backward looking investment, as well as by the licensing and construction time of the asset, lead to periods of over and under investment. This in turn leads to a cyclical long term price behavior, driven by a stochastic growth in demand, which does not settle to an equilibrium level. The structure of the problem indicates that the presence of liquid forward markets could reduce the information delay, and help stabilize the system. This assumes however that forward markets contain information which is not reflected in historical spot prices, or that is otherwise part of the public knowledge. Further research into the effect of forward markets on information flow could involve simulations of bottom up, agent-based models, to determine the extent to which locally held information is reflected in the forward price. While it may not be possible to accurately calibrate such models to the market, they would provide important qualitative insights into optimal decision rules for investors, as well as intelligent market designs for the deregulated electricity industry.

In the final part of the paper, the dynamics of spot price and new investment were linked to the physical reliability of the system. Periods of under-investment not only lead to higher price levels, but also reduce the reserve margin of available generation, and can lead to generation deficiency and blackouts. The first reaction of regulators to periods of high prices is often to try to force price levels back down through the use of price caps. Price caps, however, inhibit the economic feedback which would allow the market to readjust itself. Imposing the caps reduces the rate of new investment, leading to a slower recovery from the price hike. If the regulator continues to force the issue by reducing the cap levels, the lack of new investment will eventually lead to an erosion of the reserve margin, leading to load curtailments and blackouts in the system. The results presented in this paper indicate that regulators have to be cautious in the use of price caps. They must respect the unique characteristics of electricity as a commodity: non-storability, inelasticity of demand, and a highly constrained transmission system. These characteristics lead to an uncommonly strong link between market price signals and physical stability. Any attempt to block the true economic signals from the market could therefore prove disastrous.

Chapter 11

Conclusion

In this book we attempted to introduce a new modeling framework through which to address the uncertainty facing participants in competitive electricity markets. The lack of economic storage of electricity leads a decoupling of prices over time, creating a situation where the dimension of the uncertainty grows linearly with the time horizon of interest. Similarly, each node in the network represents a new special dimension for the market participants to consider. The escalating dimensionality of the power pricing problem threatens to render all standard approaches to financial decision making useless. Arbitrage and replicating portfolios cannot be constructed, over time or space, thus providing no guidance to the temporal or special correlation of electricity prices. To overcome this problem we introduced a model where the spot price of electricity is the output of a set of supply and demand processes. By switching the underlying states from price-based to a quantity, or bid-based, one can capture the temporal information contained in the physical and economic processes governing the demand and supply of electricity. These processes include temperature, economic growth, fuel prices, and unexpected plant failures. Joining these processes together, we build the term structure of electricity prices from its basic components. This approach has several advantages. Since competitive electricity markets are a relatively new phenomenon, there is a very limited amount of historical data available for the spot market. Furthermore, the market structure and regulatory rules are constantly evolving, so that much of the existing data is outdated. In contrast, physical processes such as temperature changes are not affected by deregulation. Indeed, given the lack of price transparency for the end consumer, the entire demand process remains largely unaffected by the regulatory changes. This allows us to draw on decades of demand and/or temperature data in calibrating the model.

In applying the bid-based model to a series of decision problems facing market participants, we found that the switch to price as an output rather than a state variable had several advantages. In the load serving problem, it was possible to reduce the complexity of the dynamic constraints by capturing the quantity risk of the ESP as a state in the price process. When addressing the issue of locational price spreads, we were able to incorporate the network flow constraints explicitly into the price model, thus projecting the nonlinear behavior of power flow onto the resulting price distributions. Finally, in examining the long term behavior of electricity prices, we were able to use the bid-based model to illustrate the delays in the feedback effects from the stochastic demand growth to the investment in new capacity. This delay has tremendous effects on price dynamics, as well as on the physical reliability of the system.

The models introduced in this book are mainly a tool for communicating the advantage of an alternative set of state variables in the price modeling process for non-storable commodities. Due to the limited amount of available data, rigorous testing against a general set of model formulations was not possible. As the markets mature, they should provide more information regarding optimal choices of model structure. In addition, the emergence of new technologies for load management on a retail level is likely to generate more elasticity in the overall market demand for electricity. As a result, further development of the model is required to cope with alternate shapes of the supply and demand curves.

When addressing the decision problems facing market participants, the focus in this book was on posing the dynamic optimization formulation, and identifying the factors which contributed to the complexity of the solution. Significant work is needed in this area in order to implement the actual optimization in an efficient manner.

Throughout the book we have considered the problem of replicating non-traded obligations with publicly traded contracts. As electricity forward and derivatives markets become more liquid, one can begin to ask the question of what would constitute a reasonable set of spanning contracts; that is, a set of contracts which can be used to dynamically replicate most over the counter and non-traded obligations. As indicated in the chapter on electricity spot and forward dynamics, strict arbitrage theory would require thousands of contracts in the spanning portfolio, given the temporal decoupling of electricity prices. However, based on the results from the bid-based model, one might argue that limited number of contracts could be used to approximate any derivative based on both temporal and special uncertainty. The problem becomes even more interesting when taking into account the possibility of a joint trading strategy in electricity and related commodities. The bid-based model can be extended to explicitly incorporate fuel prices as well as temperature levels. This would theoretically allow a user to form a replicating portfolio for electricity using oil and gas contracts, as well as weather derivatives, based on heating and cooling degree days. It further suggests that a savvy marketer may be able to take advantage of previously unrecognized model-based arbitrage between the different commodities.

Though not part of the original scope of this book, we found that the modeling presented lent some insight into the impact of market structure and government intervention on the physical and economic prosperity of the system. As shown in the chapter on long term price dynamics, the futures market serves a role, not only as a tool for hedgers and speculators, but as a medium for the transfer of information. The analysis suggests that the presence of a transparent futures market may be crucial for the physical and financial stability of the system by preventing unnecessary delays in future investment. A comparative study of the reliability levels in regions with and without futures exchanges could provide some interesting insights into the validity of this claim.

Appendix A

Derivation of Principal Components

Using Principal Component Analysis (PCA), it is possible to reduce the dimensionality of the problem by defining a new orthogonal basis. PCA generates a new orthogonal set of j variables, j ≤ n, where n is the number of the original variables in the observation set. They are called Principal Components (PC). The new variables are selected so that those describing the same driver can be replaced with a single new variable. Each principal component is a linear combination of the original variables. Because PCs are orthogonal to each other, there is no redundant information.

The first PC is a single axis in space. When each observation is projected on that axis, the resulting values form observations of a new variable, the variance of which is the maximum among all possible choices for the first axis. The second PC is orthogonal to the first, and the second variable's variance is again maximal among all possible choices for this axis. As more and more PCs are selected, they contain less and less variance. The vector, **X,** can be written as in (4), j = n = 2.

$$\mathbf{X} = a_1 y_1 + a_2 y_2 + \ldots + a_n y_n = b_1 v_1 + b_2 v_2 + \ldots + b_j v_j$$

The total number of PCs is usually equal to the number of original variables n. However, the first m PCs usually account for most of the variance in the original observations, j ≤ n. The sum of variances of the new variables equals the sum of the variances in the new variables.

$$\sum_{i=1}^{n} \operatorname{var}(v_i) = \sum_{i=1}^{n} \operatorname{var}(y_i)$$

The iterative procedure of principal component derivation can be summarized in the following steps:

1. Find the largest PC: maximize the variance of $\mathbf{b}_1^T \mathbf{v}_1 = b_{11}v_1 + \ldots + b_n v_n$:

$$\max \left[\mathrm{var}(\mathbf{b}_1^T \mathbf{x}) = \mathbf{b}_1^T \mathbf{C} \mathbf{b}_1 \right] \qquad \text{s.t.} \qquad \mathbf{b}_1^T \mathbf{b}_1 = \sum_{i=1}^{n} b_{1i}^2 = 1$$

where $\mathbf{b}_1$ is the vector of weights of the first principal component ν_1 and $\mathbf{C}$ is the covariance matrix of $\mathbf{x}$. The condition of $\mathbf{b}^T\mathbf{b} = 1$ is necessary for a unique solution to exist; otherwise the weights could become arbitrarily big, leading to infinite variance.

$$\mathbf{b}_1 = \left[b_{11}, b_{12}, \ldots, b_{1n} \right]^T$$
$$\mathbf{v}_1 = \left[v_{11}, v_{12}, \ldots, v_{1n} \right]^T$$

2. Repeat the process for the subsequent PCs, until the number of PCs = rank($\mathbf{C}$).

3. Determine how many PCs are necessary to describe the process adequately. Form the reduced-order principal component $[j \times n]$ matrix ν^*, where only the first m PCs are retained. A detailed description of the routine can be found in [22].

The eigenvalue λ_i, associated with the i-th PC corresponds to the equivalent number of variables this PC represents. A PC with an eigenvalue of $\lambda_i = 3.9$ describes on average as much variance as 3.9 original variables. Dividing the eigenvalue by the total number of PCs, j, we can obtain a total percentage on variance explained by each PC.

When all n PC have been determined, it is necessary to determine j: how many PCs are necessary to describe the data accurately enough. The three most common measures are:

1. Retain all PCs that represent more variance than original variables on average (its $\lambda_i < 1$).

2. The Scree plot. The incremental plot of variance accounted for by every PC is called the scree plot. The number of points before leveling-off of the curve is the number of PCs retained.

3. Total variance of the data accounted for by the retained PCs. Some authors propose to retain as many PCs as to account for about 90% of the

variance [23], while others propose less stringent criteria, depending on the reasons for performing the PCA [21].

Appendix B

Maximum Likelihood Estimation with Kalman Filter

Unknown parameters of the stochastic model can be estimated using Maximum Likelihood Estimation (MLE) coupled with the Kalman Filter (KF) state estimator. The iterative procedure estimates parameters of the model as to minimize the likelihood function and then computes the resulting system response using the Kalman Filter. In the appendix we derive the Kalman Filter and outline the MLE procedure.

11.1 DERIVATION OF KALMAN FILTER

A Discrete Kalman Filter is a technique for the estimation of states of a stochastic system [1]. It consists of a set of mathematical equations and provides an efficient recursive solution for the least-squares method. It addresses the problem of estimation of states **x** of a process, described by stochastic difference equation

$$\mathbf{x}_{t+1} = \mathbf{A}\mathbf{x}_t + \mathbf{B}\mathbf{u}_t + \mathbf{H}_t\boldsymbol{\eta}_t$$

where **A** is the system matrix, relating the system state $\mathbf{x}_t$ at time t to the next state $\mathbf{x}_{t+1}$at time t+1 in the absence of the controlling input. **B** is the matrix relating the input $\mathbf{u}_t$ to the state $\mathbf{x}_t$. The measured noisy system output at time t y_t is

$$y_t = \mathbf{C}\mathbf{x}_t + \varepsilon_t$$

The random variables η_t and ε_t represent process and measurement noise. They are assumed to be independent of each other and with normal probability distributions.

$$\boldsymbol{\eta}_t \approx N(0,\mathbf{Q})$$
$$\varepsilon_t \approx N(0,\mathbf{R})$$

To write the Bid-based model in state space form, the system state $\mathbf{x}_t$, the process noise η_t, the input $\mathbf{u}_t$ and output y_t signals, the system matrices **A**, **B**, **C** and Γ, and the noise covariance matrices **Q** and **R** take the following values.

$$\mathbf{x}_t = \begin{bmatrix} w_t \\ \delta_t \end{bmatrix} \qquad \boldsymbol{\eta}_k = \begin{bmatrix} z_k \\ z_k^{\delta} \end{bmatrix} \qquad y_t = [w_t] \qquad \mathbf{u}_k = \begin{bmatrix} 1 \\ 1 \end{bmatrix}$$

$$\mathbf{A} = \begin{bmatrix} 1-\alpha, \ \alpha \\ 0, \ 1 \end{bmatrix} \qquad \mathbf{B} = \begin{bmatrix} \kappa \\ \kappa \end{bmatrix} \qquad \mathbf{C} = [1,0] \qquad \mathbf{\Gamma} = \begin{bmatrix} \sigma, \sigma^{\delta} \\ 0, \sigma^{\delta} \end{bmatrix}$$

$$\mathbf{Q} = \begin{bmatrix} 1, 0 \\ 0, 1 \end{bmatrix} \qquad \mathbf{R} = 1$$

Let's define $\hat{\mathbf{x}}_{t+1/t}$ as our a-priori estimate of the state vector at step t+1, and $\tilde{\mathbf{x}}_{t+1/t+1}$ our a-posteriori estimate of the state vector at t+1, given measurement $\mathbf{z}_{t+1}$. We can then define a-priori and a-posteriori estimate errors $e_{t+1/t}$ and $e_{t/t}$ as

$$\mathbf{e}_{t+1/t} = \mathbf{x}_t - \hat{\mathbf{x}}_{t+1/t}$$
$$\mathbf{e}_{t/t} = \mathbf{x}_t - \tilde{\mathbf{x}}_{t/t}$$

An estimate of the a-priori estimate error covariance is therefore $\mathbf{P}_{t+1/t}$, with the a-posteriori estimate error covariance being $\mathbf{P}_{t/t}$.

$$\mathbf{P}_{t+1/t} = E\left[\mathbf{e}_{t+1/t}\mathbf{e}_{t+1/t}^{T}\right]$$
$$\mathbf{P}_{t/t} = E\left[\mathbf{e}_{t/t}\mathbf{e}_{t/t}^{T}\right]$$

The Kalman Filter algorithm computes the optimal a-posteriori estimate $\tilde{\mathbf{x}}_{t+1/t+1}$ as a linear combination of the a-priori estimate $\hat{\mathbf{x}}_{t+1/t}$ and a weighted difference between an actual measurement $\mathbf{z}_t$ and predicted measurement $\mathbf{C}\hat{\mathbf{x}}_{t+1/t}$. When they agree completely, the residual $\bar{y}_{t+1/t}$ is zero.

$$\tilde{\mathbf{x}}_{t+1/t+1} = \hat{\mathbf{x}}_{t+1/t} + \mathbf{K}_{t+1}\left(y_{t+1} - \mathbf{C}\hat{\mathbf{x}}_{t+1/t}\right)$$

The factor **K** in the equation is called the Kalman Gain and is chosen in such a way as to minimize the a-posteriori covariance $\mathbf{P}_{t/t}$.

11.2 KF ALGORITHM

The Kalman Filter algorithm is an iterative procedure that estimates process states as new measurements become available at each time step. Using initial estimates of system state $\mathbf{x}_{0/0}$ and a-posteriori error covariance $\mathbf{P}_{0/0}$, it computes the optimal a-posteriori estimate $\tilde{\mathbf{x}}_{t+1/t+1}$ and the pertaining Kalman Gain K. The procedure is described below:

1. Select initial estimates: $\mathbf{x}_{0/0}$ $\mathbf{P}_{0/0}$
2. Compute time update (prediction) equations:

    a-priori estimate of state vector **x**,
    $$\hat{\mathbf{x}}_{t+1/t} = \mathbf{A}\tilde{\mathbf{x}}_{t/t} + \mathbf{B}\mathbf{u}_t$$

    a-priori error covariance matrix,
    $$\mathbf{P}_{t+1/t} = \mathbf{A}\mathbf{P}_{t/t}\mathbf{A}^T + \mathbf{\Gamma}\mathbf{Q}\mathbf{\Gamma}^T$$

3. Compute measurement update (correction) equation:

    Kalman gain,

$$\mathbf{K}_{t+1} = \mathbf{P}_{t+1/t}\mathbf{C}_{t+1}^{T}\left[\mathbf{C}_{t+1}\mathbf{P}_{t+1/t}\mathbf{C}_{t+1}^{T}\right]^{-1}$$

residual measurement innovation

$$\bar{y}_{t+1/t} = y_{t+1} - \mathbf{C}\hat{\mathbf{x}}_{t+1/t}$$

a-posteriori estimate of state vector $\mathbf{x}$

$$\tilde{\mathbf{x}}_{t+1/t+1} = \hat{\mathbf{x}}_{t+1/t} + \mathbf{K}_{t+1}\bar{y}_{t+1/t}$$

a-posteriori error covariance estimate

$$\mathbf{P}_{t+1/t+1} = \left[\mathbf{I} - \mathbf{K}_{t+1}\mathbf{C}\right]\mathbf{P}_{t+1/t}$$

4. Repeat 2 and 3 for all $t \in [1,...T]$

   The procedure is schematically shown in figure b1.

11.3 MAXIMUM LIKELIHOOD ESTIMATION OF MODEL PARAMETERS

The idea behind Maximum Likelihood Estimation is to compute the optimal parameters of the model by iteratively modifying them to minimize a likelihood function [24], [25], [26].

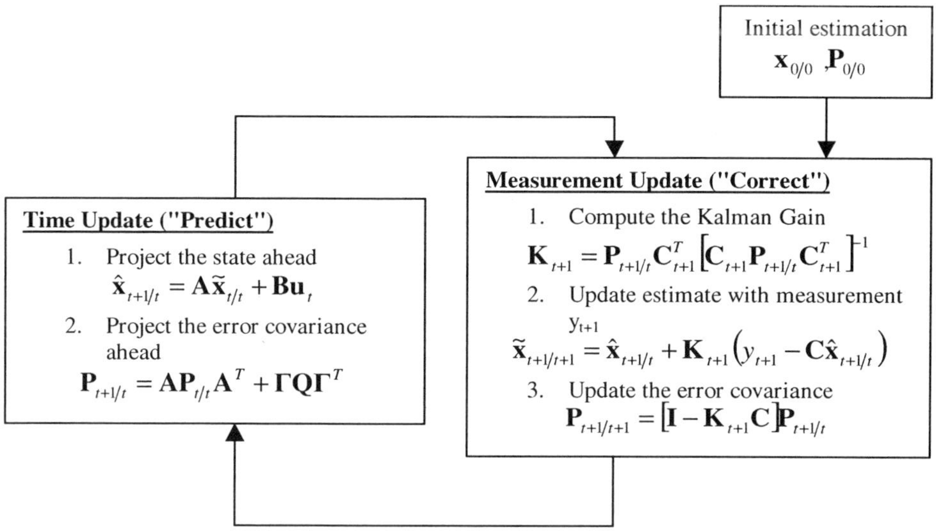

Fig. B.1 Kalman Filter operation flowchart.

After constructing the model representation in the state space and setting up the KF procedure, we construct a vector of unknown parameters θ that contains the unknown parameters of the model.

$$\boldsymbol{\theta} = \left[\alpha \quad \kappa \quad \sigma \quad \sigma^{\delta}\right]^T$$

Using the covariance of the innovation process $\mathbf{N}_{t+1/t}$, obtained by the Kalman Filter,

$$\mathbf{N}_{t+1/t} = \mathbf{C}\mathbf{P}_{t+1/t}\mathbf{C}^T$$

we can construct a log likelihood function J.

$$J = \log L = -\frac{1}{2}\sum_{t=1}^{T-1}\left[\bar{y}_{t+1/t}^T\mathbf{N}_{t+1/t}^{-1}\bar{y}_{t+1/t} + \log\left(\det\left(\mathbf{N}_{t+1/t}\right)\right)\right]$$

The procedure iteratively updates the parameter vector θ according to the equation

$$\boldsymbol{\theta}^{i+1} = \boldsymbol{\theta}^i - \rho^i\mathbf{M}^{-1}\left(\boldsymbol{\theta}^i\right)\frac{\partial J\left(\boldsymbol{\theta}^i\right)}{\partial\boldsymbol{\theta}}$$

where **M**(a) is a Hessian matrix of the log likelihood function

$$\mathbf{M}(\mathbf{a}) = \frac{\partial^2 J(\mathbf{a})}{\partial a_i \partial a_j}$$

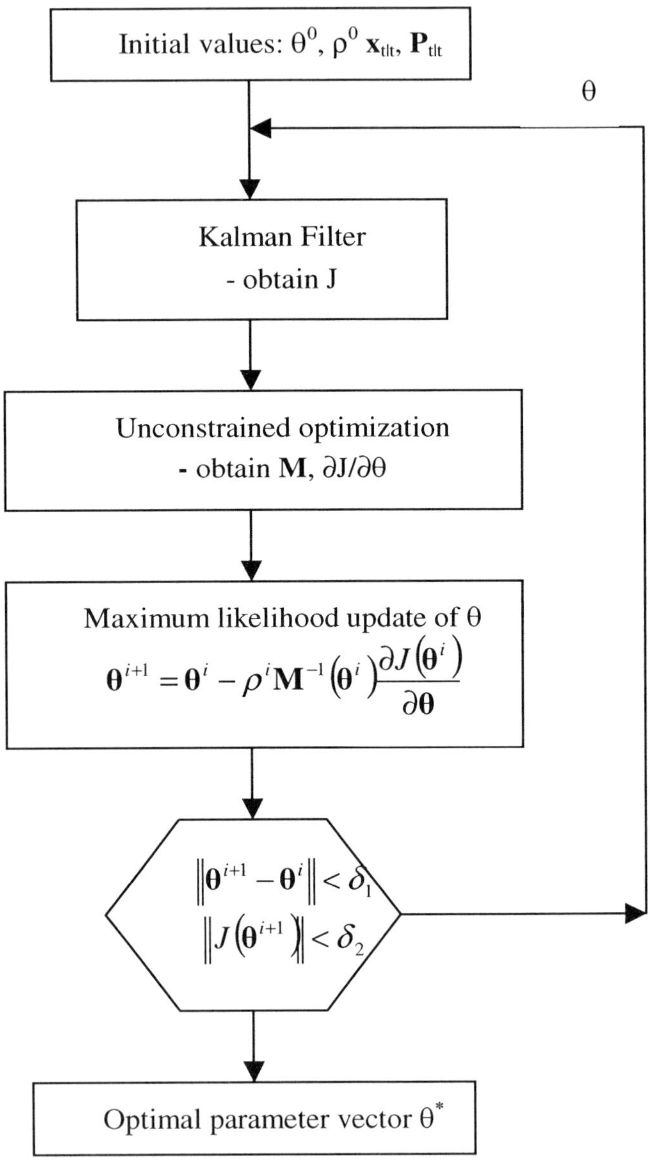

Fig. B.2 Maximum Likelihood estimation flowchart

References

[1] Emanuel Derman, "Model Risk", Quantitative Strategies Research Notes, Goldman Sachs, April 1996

[2] Hull, J.: Options, Futures and other Derivatives, 4th ed., Prentice-Hall, 1999.

[3] F. Black and M. Scholes, "The Pricing of Options and Corporate Liabilities," Jounal of Political Economy, 81 May-June 1973, 637-659.

[4] R.C. Merton, "Theory of Rational Option Pricing," Bell Jounal of Economics and Management Science, 4 (Spring 1973), 141-183.

[5] Schwartz, E.: "The Stochastic Behavior of Commodity Prices: Implications for Valuation and Hedging", *The Journal of Finance,* Vol. 7, No. 3, July 1997.

[6] Dixit, A.K., Pindyck, R.S., Investment Under Uncertainty, Princeton University Press, 1994.

[7] Allen, E., Ilic, M.D., Price-Based Commitment Decisions in the Electricity Market, Springer-Verlag, London, 1999.

[8] Joy C., "Pricing, Modeling and Managing Physical Power Derivatives", Energy Modeling and the Management of Uncertainty, Risk Publications, 1999.

[9] Baker, M. P., Mayfield, E. S., Parsons, J. E.: "Alternative Models of Uncertain Commodity Prices for Use with Modern Asset Pricing Methods", *The Energy Journal*, Vol. 19, No. 1.

[10] Deng, S.: "Stochastic Models of Energy Commodity Prices and Their Applications: Mean-reversion with Jumps and Spikes", POWER report PWP-073, February 2000, Berkley, California.

[11] Rudkevich, A., Duckworth, M., Rosen, R.; "Modeling Electricity Pricing in a Deregulated Generation Industry: The Potential for Oligopoly Pricing in a Poolco", *The Energy Journal*, Vol. 19, No. 3.

[12] Baldick R., Gant R., Kahn E., "Linear Supply Function Equilibrium: Generalizations, and Limitations", POWER report PWP-078, August 2000, Berkley, California.

[13] Hobbs, B.F., Metzler, C.B., Pang, J.S., "Strategic Gaming Analysis for Electric Power Systems: An MPEC Approach", IEEE Trans. on Power Systems, Vol. 15, No. 2, May 2000, pp. 638-645.

[14] Rudkevich, A., "Supply Equilibrium in Poolco Type Power Markets: Learning All the Way", Draft, March 5, 1999.

[15] Visudhiphan, P., Ilić, M.D., "Dynamic Game-based Modeling of Electricity Markets", *1999 IEEE PES Winter Power Meeting*, New York, New York City, February 1999.

[16] Visudhiphan, P., Ilić, M.D., "Dependence of Generation Market Power on the Demand/Supply Ratio: Analysis and Modeling", *2000 IEEE PES Winter Power Meeting*, Singapore, January 2000.

[17] Backerman, S. R., Denton, M. J., Rassenti, S. J., Smith, V. L., "Market Power in a Deregulated Electrical Industry: An Experimental Study", Economic Science Laboratory, University of Arizona, Tuscon, AR.

[18] Skantze, P., Chapman, J., Ilić, M.D., "Stochastic Modeling of Electric Power Prices in a Multi-Market Environment", *Transactions of IEEE PES Winter Meeting*, Singapore, January 2000.

[19] Skantze, P., Gubina, A., Ilic, M., "Bid-based Stochastic Model for Electricity Prices: The Impact of Fundamental Drivers on Market Dynamics", MIT Energy Laboratory Technical Report EL 00-004, November 2000.

[20] Kosecki, R., "Fuel-based Power Price Modeling", Energy Modeling and the Management of Uncertainty, Risk Publications, 1999.

[21] G. H. Dunteman, Introduction to Multivariate Statistics. Sage Publications; Beverly Hills, 1984.

[22] G.H. Dunteman, Principal Component Analysis. Sage Publications; Newbury Park, 1989.

[23] B. Flury, Common Principal Component and related Multivariate Models. Wiley; New York, 1988, ISBN 0-471-63427-1

[24] Bolland, P.J., Connor, J.T.: "A Constrained Neural Network Kalman Filter for Price Estimation in High Frequency Financial

Data", *Int. Journal of Neural Systems*, Vol. 8, No. 4, August 1997, pp. 399-415.

[25] Burmeister, E., Wall, K.D.: "Kalman Filtering Estimation of Unobserved Rational Expectations with an Application to the German Hyperinflation"; *Journal of Econometrics*, Vol. 20, North-Holland Publ. 1982, pp. 255-284.

[26] Manoliu, M., Tompaidis, S.: "Energy Futures Prices: Term Structure Models with Kalman Filter Estimation", White Paper, Publication accessible at www.caminus.com.

[27] S. K. Kachigan, Statistical Analysis – An Interdisciplinary Introduction to Univariate & Multivariate Methods. Radius Press, New York, 1986.

[28] Chui, C.K., Chen, G.: Kalman Filtering, with Real-Time Applications; Springer, New York, 1999, ISBN 3-540-64611-6.

[29] Welch, G., Bishop, G.: "An Introduction to the Kalman Filter", UNC-Chapel Hill, Working Paper TR 95-041, June 6, 2000.

[30] Mehra, R.K., "On the Identification of Variances and Adaptive Kalman Filtering"; *IEEE Trans. on Automatic Control*, Vol. AC-15, No. 2, April 1970, pp. 175-184.

[31] Mehra, R.K.: "Approaches to Adaptive Filtering", *IEEE Trans. on Automatic Control*, Vol. AC-17, No. 5, Oct. 1972, pp. 693-8.

[32] Matlab Statistics Toolbox User's Guide, The MathWorks, Inc., 1999.

[33] Fumagalli, E., Black, J., Ilic, M.D., Vogelsang, I., "A Reliability Insurance Scheme for the Electricity Distribution Grid", MIT Energy Laboratory Working Paper EL 01-001, January 2001.

[34] Joskow, P.L., "Why do We Need Electricity Retailers?" MIT-CEEPR Working Paper 2000-01 January 2000.

[35] Ilić, M. D., Skantze, P., "Electric Power Systems Operation by Decision and Control: The Case Revisited", *IEEE Control Systems Magazine*, Vol. 20, No. 4, August 2000.

[36] Skantze, P., Ilić, M.D., "The Joint Dynamics of Electricity Spot and Forward markets: Implications on Formulating Dynamic Hedging

Strategies", MIT Energy Laboratory Technical Report EL 00-005, MIT, November 2000.

[37] Baykal-Gursoy, M., Ross, K.W., "Variability Sensitive Markov Decision Processes", Mathematics of Operations Research, Vol 17, No. 3. August 1992.

[38] Filar, J.A., Kallenberg, L.C.M., Lee, H.-M., "Variance-Penalized Markov Decision Processes", Mathematics of Operations Research, Vol 14, No. 1. Febuary 1989.

[39] White, D.J., "Dynamic Programming and Probabilistic Constraints", Manchester, England, August 24, 1971.

[40] Bertsekas, D.P., Tsitsiklis, J.N., Neuro-Dynamic Programming, Athena Scientific, 1996.

[41] Bertsekas, D. P., Dynamic Programming: Deterministic and Stochastic Models, Prentice-Hall, 1987.

[42] Puterman, M.L., Markov Decision Processes: Discrete Stochasic Dynamic Programming, John Wiley & Sons, 1994.

[43] Wagner, M., "Hedging Optimization Algorithms for Deregulated Electricity Markets", upcoming masters thesis at the Massachusetts Institute of Technology, June 2001.

[44] Deng, S., Johnson B., Sogomonian, A., (1998) 'Exotic Electricity Options and the Valuation of Electricity Generation and Transmission Assets,' presented at the Chicago Risk Management Conference, Chicago, May 1998.

[45] Tseng, C-L., Barz, G., (1998), 'Short-Term Generation Asset Valuation,' Proceedings of the Thirty-second Annual Hawaii International Conference on System Sciences,

[46] Skantze, P., Visudhiphan, P., Ilić, M.D., "Valuation of Generation Assets with Unit Commitment Constraints under Uncertain Fuel Prices", MIT Energy Laboratory Technical Report EL 00-006, MIT, November 2000.

[47] Ilić, M., Zaborszky, J., Dynamics and Control of Large Electric Power Systems, Wiley & Sons, 2000, ISBN 0-471-29858-1.

[48] Gubina F., Grgič D., Banič I., "A Method for Determining the Generators' Share in a Consumer Load ", *to appear in IEEE T-PWRS in 2001.*

[49] Joskow, P.L., Tirole, J., "Transmission Rights and Market Power on Electric Power Networks", Rand Journal of Economics, Vol. 31, No. 3, 2000, p. 450-487.

[50] California ISO Corp., "FERC Electric Tariff", First Rep. Vol., No. I, October 13, 2000.

[51] Hogan, W., "Contract Networks for Electric Power transmission: Technical Reference", Harvard University, September 1990.

[52] Ilić, M., Hyman, L., Allen, E., Younes, Z., "Transmission Scarcity: Who Pays?", The Electricity Journal, July 1997, p. 38-49.

[53] Oren, S., Spiller, P., Varaiya, P., Wu, F., "Nodal Prices and Transmission Rights: a Critical Appraisal", University of California, Berkeley, December 1994.

[54] Yoon, Y.T., Arce, J.R., Collison, K.K., Ilić, M., "Implementation of Cluster-based Congestion Management Systems", ICPSOP 2000, "Restructuring the Power Industry for the Year 2000 and Beyond", 2000.

[55] Yu, C-N., Ilic, M.D., "Congestion Cluster-Based Markets for Transmission Management", IEEE Winter Meeting, January 1999, pp 821-832.

[56] Ott, A., "Fixed Transmission Right Auction", IEEE 1999

[57] PJM, "FTR Auction Training", April 1999, http://www.pjm.com/training/training_index.html

[58] Chicago Mercantile Exchange, Futures and Options Listing Schedule, http://www.cme.com/clearing/listings/index.html

[59] Baker, M. P., Mayfield, E. S., Parsons, J. E.: "Alternative Models of Uncertain Commodity Prices for Use with Modern Asset Pricing Methods", *The Energy Journal*, Vol. 19, No. 1.

[60] Mount, T.: "Market Power and Price Volatility in Restructured Markets for Electricity", *Proceedings of the Hawaii International Conference on System Sciences*, January 1999, Maui, Hawaii.

[61] Backerman, S. R., Denton, M. J., Rassenti, S. J., Smith, V. L., "Market Power in a Deregulated Electrical Industry: An Experimental Study", Economic Science Laboratory, University of Arizona, Tuscon, AR.

[62] Borenstein S., Bushnell J., Wolak, F., "Diagnosing Market Power in California's Restructured Wholesale Electricity Market", August 2000.

[63] Moerch von der Fehr N.-H. , Harbord D., "Spot Market Competition in the UK Electricity Industry", Economic Journal, No. 103, May 1993, pp. 531-546. Blackwell Publishers, Cambridge, MA.

[64] Mount, T.: "Market Power and Price Volatility in Restructured Markets for Electricity", *Proceedings of the Hawaii International Conference on System Sciences*, January 1999, Maui, Hawaii.

Index

agent-based modeling, vii, 58
arbitrage
  cash and carry, 20, 24, 41, 43, 45
  definition of, 18
battery, 48
bid-based model, 80, 111, 113, 114, 122, 132, 158, 166, 168, 194, 195
bidding strategy, 139, 146
Black Scholes model, 21
commodities
  non-storable, xi, 194
  storable, 3, 19, 26, 48, 55
congestion clusters, 150
congestion management, 33, 150
cross bidding, 160
double auction, 61
dynamic programming, 46, 53, 125, 126, 135
dynamic replication, 40, 169
economic equilibrium models, vii, 58
economic feedback, 190, 191
end state problem, 129
energy service provider (ESP), 32, 113, 115, 116, 117, 131, 152
experimental models, vii, 59
forward markets
  information content, 185
fundamental modeling, vii, 59
futures markets, 183, 184
generation assets, 31, 53, 72, 135, 136, 145, 150, 179, 186
Geometric Brownian Motion (GBM), 21
hedging
  dynamic, 114, 119, 165, 169
  static, 114
industrial customers, 32
investment
  dynamics of, 55, 180, 182, 184
Kirchhoff's laws, 2, 5
least squares estimation, 81, 88
load serving entity (LSE), 113
locational pricing, 5
lookup table, 136, 140, 142, 143, 144, 146
Monte Carlo techniques, 143, 144, 145
net present value (NPV), 7, 8
New York Mercantile Exchange (NYMEX), 36
options, xi, 5, 8, 12, 32, 34, 39, 152, 158, 168
  over the counter (OTC), 34, 35, 195
  swing, 39
physical risk, 153, 154
price caps, 187, 188, 189, 191
pricing
  arbitrage pricing theory (APT), xi, 7, 17, 19, 54
  risk neutral, 22

principal component analysis (PCA), 74, 83, 111
production based models, vii
reliability, 6, 113, 116, 154, 177, 179, 186, 190, 194, 195
  dynamic notion of, x, 185
retail customers, 113, 115
risk measures
  mean-variance formulation, 16, 114, 124
  value at risk (VAR), 9, 10, 129, 131
risk neutral pricing, 22
risk preference, 8, 9, 10, 13, 14, 17, 114, 117, 124
scheduled maintenance, 62
spark spread, 39
storage
  cost of, 3, 20, 46
  gas, 44, 45
  hydro-electric, vii, 46
  pump storage facilities, 46, 48
  strategies, vii, 43
supply bid curve, 72, 74, 81
temperature, xii, 4, 39, 55, 60, 122, 193, 195
term one, 29
time scale separation, 55, 70
transmission assets
  valuation, 149
transmission rights, 5, 39, 151, 152, 153, 154, 155, 158, 161, 162, 165, 170, 173
  fixed, 151, 154
  flexible, 151, 154, 155
unit commitment, 46, 73, 135, 136, 137, 139, 141, 142, 145
valuation, vi, ix, xi, xii, 1, 7, 12, 13, 16, 69, 135, 136, 139, 143, 144, 146, 149, 151, 157, 161
  market based, 149
  of generation assets, 136
weather, 2, 4, 24, 39, 61, 68, 195
Wiener process, 5, 21, 27, 157, 165
zonal pricing, 150